Tunable RF Components and Circuits

APPLICATIONS IN MOBILE HANDSETS

Devices, Circuits, and Systems

Series Editor
Krzysztof Iniewski
CMOS Emerging Technologies Inc., Vancouver, British Columbia, Canada

Atomic Nanoscale Technology in the Nuclear Industry
Taeho Woo

Nano-Semiconductors: Devices and Technology
Krzysztof Iniewski

Electrical Solitons: Theory, Design, and Applications
David Ricketts and Donhee Ham

Radiation Effects in Semiconductors
Krzysztof Iniewski

Electronics for Radiation Detection
Krzysztof Iniewski

Semiconductor Radiation Detection Systems
Krzysztof Iniewski

Internet Networks: Wired, Wireless, and Optical Technologies
Krzysztof Iniewski

Integrated Microsystems: Electronics, Photonics, and Biotechnology
Krzysztof Iniewski

FORTHCOMING

Telecommunication Networks
Eugenio Iannone

Optical, Acoustic, Magnetic, and Mechanical Sensor Technologies
Krzysztof Iniewski

Biological and Medical Sensor Technologies
Krzysztof Iniewski

Tunable RF Components and Circuits

APPLICATIONS IN MOBILE HANDSETS

EDITED BY

Jeffrey L. Hilbert

WiSpry, Incorporated, Irvine, California, USA

Krzysztof Iniewski

MANAGING EDITOR

CMOS Emerging Technologies Research Inc.
Vancouver, British Columbia, Canada

CRC Press
Taylor & Francis Group
Boca Raton London New York

CRC Press is an imprint of the
Taylor & Francis Group, an **informa** business

MATLAB® is a trademark of The MathWorks, Inc. and is used with permission. The MathWorks does not warrant the accuracy of the text or exercises in this book. This book's use or discussion of MAT-LAB® software or related products does not constitute endorsement or sponsorship by The MathWorks of a particular pedagogical approach or particular use of the MATLAB® software.

CRC Press
Taylor & Francis Group
6000 Broken Sound Parkway NW, Suite 300
Boca Raton, FL 33487-2742

First issued in paperback 2020

© 2016 by Taylor & Francis Group, LLC
CRC Press is an imprint of Taylor & Francis Group, an Informa business

No claim to original U.S. Government works

ISBN-13: 978-1-4987-1889-9 (hbk)
ISBN-13: 978-0-367-65627-0 (pbk)

Visit the Taylor & Francis Web site at
http://www.taylorandfrancis.com

and the CRC Press Web site at
http://www.crcpress.com

*For Cyndie, who makes everything possible
and worthwhile, and who proves every day
that dreams can come true*

Contents

Preface

An invasion of armies can be resisted, but not an idea whose time has come.

Victor Hugo (1852)

Tunable RF's time has come. Arguably, it may be long overdue. With the continuing explosive growth of mobile communications on a global scale, whether measured in terms of users, handsets, data traffic, networks, and/or frequency bands, ubiquitous, reliable wireless connectivity has never been more important. But innovation in the portion of the mobile handset architecture that most directly enables network connectivity, the radio frequency (RF) front end, has taken a backseat to continuing advancements in other parts of the handset. Whereas most mobile handset users have some idea about the functions and value of a faster processor, more memory, or a larger screen since these components are more easily connected to observable user benefits, the same is not true about the value or function of the antennas, amplifiers, switches, and filters in the RF front end. (Witness the "Antennagate" episode a few years ago, and the resulting calls I received from various reporters, some of whom wanted to know what an antenna does and why it is so important.) Of course, without a robust front-end solution, accessing the benefits provided by advancements in other parts of the handset becomes, at a minimum, problematic. And it is perhaps interesting to note that, when surveyed, a majority of handset users regularly cite battery life, dropped calls, and data speed as among the biggest concerns with their mobile service, all of which are heavily driven by the performance of the RF front end.

So, why have advancements in the RF front-end architecture lagged those in the rest of the handset? Perhaps the simplest answer is, until recently, complementary metal oxide semiconductor (CMOS) technologies and design techniques have not supported the levels of performance required of RF components. As a result, a collection of various higher-performance specialty technologies (acoustics, ceramics, magnetics, gallium arsenide semiconductors) have been utilized to implement the various RF functions. These functions have been interconnected into chains of RF components through 50 ohm interfaces, with each such chain optimized to work for a specific mode or frequency of operation. As more modes and frequencies are added to the handset, more chains are added, and interconnected by higher and higher throw count switches. A second answer is that by throwing more processing power and memory into the handset, and shifting more hardware functionality into software, in other words, by using Moore's law scaling, we have been able to (barely) keep pace with demand. However, with no end to data demand in sight, and network performance improvements already defined for future releases of 4G, we have reached a tipping point where traditional RF front-end architectures have become a major bottleneck in enabling the industrial design and performance attributes of handsets that customers are demanding.

At first glance, in the context of the handset RF front end, the words "tunable" and "RF" may not seem to belong together. We define a tunable RF function to be one whose target frequency of operation and performance parameters are dynamically adjustable under software control. Such adjustments are typically in response to changes in the micro- and/or macro-environment of the handset. Current front-end architectures are generally "switchable" at the RF chain level (as discussed earlier) or between instances of components within a chain and as such, can broadly be considered to be tunable but at a lower level of precision than possible with individual RF components or modules. We will discuss both levels of "tunability" within this book along with the substantial progress toward future RF front-end architectures that are completely tunable without the use of switches at the RF chain level ("switchless" architectures).

It is difficult to establish the exact start of today's tunable RF market; however, our company, WiSpry, along with Paratek Microwave (later purchased by Research in Motion), was engaged in working with handset original equipment manufacturer (OEM) customers on tunable impedance matching for antennas beginning about the better part of 10 years ago. Since then, the market has grown rapidly, and today it would be difficult to find a smartphone that does not employ at least one tunable RF function implemented using either electrostatic microelectromechanical systems (MEMS), barium strontium titanate (BST), silicon-on-insulator (SOI) field effect transistors (FETs), or, for some tunable functions, high-performance RF CMOS technology.

This book serves as a snapshot of the state of the art in tunable RF circa 2015. The book is written by leading practitioners in the field, and between us, we comprise a majority of today's commercial market share for tunable devices in mobile handset applications. The goal of the book is to provide a technical introduction to the field and to document the foundational work that has been done to date. Chapter 1 serves as an introduction and provides an overview of the tunable RF market along with a glimpse into the future. Chapters 2 and 3, contributed by Qorvo (Greensboro, North Carolina), ON Semiconductor (Nashua, New Hampshire) and BlackBerry (Waterloo, Ontario, Canada), focus on RFSOI technology and BST technology, respectively, along with an introduction to the applications of these technologies to tunable functions. Chapters 4 and 5, contributed by Skycross (Viera, Florida) and Ethertronics (San Diego, California), discuss the applications of tuning to mobile handset antenna structures and systems. Beginning with Chapter 6, the next three chapters (contributed by Cavendish Kinetics [San Jose, California], Peregrine Semiconductor [San Diego, California], and ON Semiconductor [Nashua, New Hampshire]) dive into antenna tuning applications utilizing RF-MEMS, SOI, and BST technologies, respectively. Chapter 9 concludes the detailed discussion of antenna tuning by offering a perspective on the topic from a handset OEM, Huawei (San Diego, California). In Chapter 10, Nujira (Cambridge, United Kingdom) provides an in-depth discussion of power amplifier envelope tracking, a rapidly emerging and important technique for improving efficiency. In Chapter 11, DelfMEMS (Villenuve D'Ascq, France) discusses using RF-MEMS switches for a next-generation implementation of a switchable RF front end. A case study of tunable radio architectures by Interdigital (Melville, New York) is presented in Chapter 12, while Chapter 13 provides AT&T's (Atlanta, Georgia) network operator perspective on the evolution of the handset front end. The book concludes with a chapter from LitePoint (Sunnyvale, California) on production testing of wireless devices in the face of the continuing evolution of handset architectures and the increasing drive toward shifting such testing from a conducted to a radiated (over-the-air) performance basis.

Some readers may note a substantial portion of the book is devoted to antennas and antenna tuning. This is by design, as the antenna tuning application dominates the commercial market for tunable RF today. The rate of adoption of tunable RF and the evolution of RF front-end architectures in mobile handsets continue to accelerate so there is little doubt that future books on the topic will present a more diverse and, perhaps arguably, a more balanced overview of a then more mature market.

Over my almost 40-year career in the semiconductor industry, I have been lucky enough to contribute to a number of advances in technology, design, and design tools. In the 1970s, it was CMOS technology and the start of commercial design tools. In the 1980s, it was gate array and standard cell ASICs as both a supplier and a user. In the early 1990s, it was systems-on-a-chip (SOC), followed in the late 1990s and early 2000s by the mainstreaming of MEMS technology. Then, since about 2005, I have participated in developing the foundation of tunable RF. As we move toward a world in which always-on connectivity and on-demand access to information is available to everyone on a global scale, it seems possible that sensing, computing, and wireless communications technologies will become so pervasive as to become invisible in the fabric of society, perhaps as a step before they become biologically integrated within us. I hope my luck holds and I get another chance to play. If not, I will hope to find some small level of satisfaction at having played a small role in bringing about (with apologies for paraphrasing a portion of a Winston Churchill quote) "the end of the beginning" of something that ultimately should be wonderful for all.

Jeffrey L. Hilbert
Dana Point, California

MATLAB® is a registered trademark of The MathWorks, Inc. For product information, please contact:

The MathWorks, Inc.
3 Apple Hill Drive
Natick, MA 01760-2098 USA
Tel: 508-647-7000
Fax: 508-647-7001
E-mail: info@mathworks.com
Web: www.mathworks.com

Acknowledgments

I thank all of the contributors to this book for their participation and the many hours they devoted. Many of us are direct competitors in the market, so their willingness to play a role is greatly appreciated along with their contributions to tunable RF. My thanks to the entire WiSpry team for their efforts over the past 12 plus years and the determination that led to the first commercial deployment of RF MEMS in consumer electronics in 2011. I also thank Kris Iniewski for inviting me to give talks at his CMOS Emerging Technologies Research Conference series, which led to the opportunity to develop this book, and to the team at Taylor & Francis for their support in the completion of the book. And finally, to my wife, Cyndie, who had to put up with the many nights and weekends I spent writing and editing.

Series Editor

Krzysztof (Kris) Iniewski manages R&D at Redlen Technologies Inc., a start-up company in Vancouver, Canada. Redlen's revolutionary production process for advanced semiconductor materials enables a new generation of more accurate, all-digital, radiation-based imaging solutions. Dr. Iniewski is also a president of CMOS Emerging Technologies Research Inc. (www.cmosetr.com), an organization of high-tech events covering communications, microsystems, optoelectronics, and sensors. In his career, Dr. Iniewski has held numerous faculty and management positions at the University of Toronto, the University of Alberta, Simon Fraser University, and PMC-Sierra Inc. He has published over 100 research papers in international journals and conferences. He holds 18 international patents granted in the United States, Canada, France, Germany, and Japan. He is a frequent invited speaker and has consulted for multiple organizations internationally. Dr. Iniewski has written and edited several books for CRC Press/Taylor & Francis, Cambridge University Press, IEEE Press, Wiley, McGraw-Hill, Artech House, and Springer. His personal goal is to contribute to healthy living and sustainability through innovative engineering solutions. He can be reached at: kris.iniewski@gmail.com.

Editor

Jeffrey L. Hilbert is the president and founder of WiSpry, Inc., a fabless semiconductor company utilizing CMOS-integrated radio frequency microelectromechanical system (RF-MEMS) technology to develop tunable RF products for the cellular communications and wireless consumer electronics markets. WiSpry defined and pioneered the rapidly growing tunable RF market segment. Hilbert has more than 37 years of executive management and technical experience in a number of leading semiconductor and MEMS companies, including LSI Logic, Compass Design Automation, AMCC, Motorola, Harris, and Coventor. Early in his career, Hilbert did pioneering work in CMOS technology and in IC design tools leading to today's design automation tools that are supplied by companies such as Cadence. As an experienced entrepreneur, he has raised over $120 million in financing to fund two consecutive start-up semiconductor companies over the past 15 years. He also has board of director and advisory board experience in the commercial, government, and academic arenas. Hilbert holds a BS in chemical engineering from the University of Florida, an MS in computer science from the Florida Institute of Technology, and has done course work toward a PhD in computer engineering from North Carolina State University.

Contributors

Rob Brownstein
Litepoint
Sunnyvale, California

Frank Caimi
Skycross
Viera, Florida

Julio Costa
Qorvo
Greensboro, North Carolina

Alpaslan Demir
Interdigital
Melville, New York

Laurent Desclos
Ethertronics
San Diego, California

Tanbir Haque
Interdigital
Melville, New York

Jeremy Hendy
Nujira
Cambridge, United Kingdom

Minh-Chau Huynh
Litepoint
Sunnyvale, California

David W. Laks
ON Semiconductor
Burlington, Ontario, Canada

Igor Lalicevic
DelfMEMS
Villenuve D'Ascq, France

Yuang Lou
AT&T Network Architectures,
    Radio Access and Devices
Atlanta, Georgia

Paul McIntosh
ON Semiconductor
Nashua, New Hampshire

Larry Morrell
Cavendish Kinetics
San Jose, California

James G. Oakes
BlackBerry
Waterloo, Ontario, Canada

Tero Ranta
Peregrine Semiconductor
San Diego, California

Sebastian Rowson
Ethertronics
San Diego, California

Jeff Shamblin
Ethertronics
San Diego, California

Ping Shi
Huawei Device USA
San Diego, California

Paul Tornatta
Cavendish Kinetics
San Jose, California

Gerard Wimpenny
Nujira
Cambridge, United Kingdom

1 Tunable RF Market Overview

Jeffrey L. Hilbert

CONTENTS

1.1 INTRODUCTION

Mobile (wireless) communications has rapidly become embedded in the fabric of society enabling and changing the ways in which we communicate in the broadest sense. Cutting across geographic and political boundaries, age, differences in ethnicity and religion, and independent of sharing and exchanging information one-to-one, one-to-many, or many-to-many, the mobile or cellular (cell) phone has become a ubiquitous personal appliance on a global scale.

It is difficult to imagine a world without mobile phones yet it has only been about 42 years since the first mobile phone call was placed by Martin Cooper of Motorola in New York City in April of 1973. About 10 years later, Motorola began selling the Dynatac 8000x. With an initial price of $3995, this phone offered mobile, analog voice communications with up to 30 minutes of talk time in a form factor of only 13 in. × 1.75 in. × 3.5 in. (Figure 1.1).

From these humble beginnings, the mobile communications market has exploded to change the way in which we live. At the same time, the mobile device has continued to rapidly grow in capabilities, performance, and talk time while shrinking in size and weight. Figure 1.2 shows a current generation, 4G mobile phone, the LG G3. This phone provides up to 21 hours of talk time in a form factor of 5.42 in. × 2.74 in. × 0.41 in., not to mention almost unbelievable improvements in functionality, performance, and ease of use. The contrast with the Dynatac 8000x is apparent.

In this chapter, we begin with a review of the status and projections for the mobile handset market. As mobile phone form factors continue to evolve in response to consumer demands, and performance and capabilities continue to grow with the use

FIGURE 1.1   Motorola Dynatac 8000x.

FIGURE 1.2   LG G3 (4G handset).

of succeeding generations of semiconductor technologies, we next turn our attention to the architecture of the mobile phone and in particular, to the radio frequency (RF) front end. Unlike other portions of the architecture, the RF front end (RFFE) has been relatively slow to evolve creating a bottleneck for future improvements and an imperative for change. Tunable RF provides a new approach to implementing the RFFE and the next three sections of this chapter introduce the tunable RF concept, summarize the current tunable RF market and applications, and provide a projection of the future impact and opportunity for tunable RF components and circuits. The final section of the chapter provides a summary of the materials for the reader's consideration.

1.2 THE MOBILE HANDSET MARKET

Just as consumer electronics has become the growth engine for the overall semiconductor industry, the mobile handset market has become the driver for growth in consumer electronics.

Since the 1990s, cellular wireless technology has advanced from second-generation (2G and 2.5G, or GSM), to third-generation (3G) technology, and then on to 3.5G, 3.9G and now, 4G (Long Term Evolution [LTE] and LTE-A [advanced]) technology. Despite multiple future, yet-to-be-released generations of 4G technology that are already defined, talks are well underway for the initial definition of 5G technologies with a target deployment date around 2020. At the same time as cellular standards have continued to advance bringing new modulation schemes, increased capacity, higher speeds, better interoperation, and improved hand-offs and backhaul, the types and volumes of information being transferred have evolved radically. From a portable replacement for the fixed, analog landline for voice communications, the network has evolved into a digital communications system initially adding support for data over an analog network, then to a data-centric architecture supporting high-speed downloads of streaming video, and now on to supporting voice over data (VoLTE). Additional frequencies have been deployed to handle the radical increase in data volumes, which are projected to increase by 61% on a compound annual growth rate (CAGR) between 2013 and 2018, reaching 15.9 EB of data per month by 2018.[1] Not only has the range of supported frequency bands for cellular communications grown both toward lower and higher frequencies, but there has also been an increasing amount of reallocation (re-farming) of existing frequency bands in the continuing search to create additional capacity.

The structure of the handset market has changed in parallel with the evolution of the cellular network. From a single type of analog phone (think the Dynatac 8000x), we moved to a market model of basic and more-advanced, feature phones. The advent of 3G networks saw the birth of the smartphone platform as the premium product offering from multiple handset original equipment manufacturers (OEMs). As time has passed, the basic phone has become the (very) low cost, entry-level phone often servicing the majority of users in the developing world, but providing the level of functionality previously provided by feature phones. The smartphone market bifurcated into basic smartphones, which have cannibalized the feature phone market, and high-end smartphones, which are now often referred to as advanced smartphones or global smartphones. Currently, the handset market can be thought of as being composed of three segments: (1) low cost (feature) phones; (2) smartphones (advanced 3G and basic 4G); and (3) advanced smartphones (4G multiregional and globally enabled models). The general trend in the number of models of a phone and the number of new phone designs per year has in recent years been decreasing as industry consolidation continues, more emphasis is placed on hardware platform-based designs, and coverage evolves from carrier- and geography-specific, to regional, super-regional, and global roaming capabilities. The corresponding reduction in the number of stockkeeping units (SKUs) as inspired by Apple's success with this approach for the iPhone has been beneficial to handset OEMs and had a substantial impact on time to market and bill-of-materials (BOM) costs.

Mobile handsets seem to have become a universal presence. But, just how pervasive has mobile communications become? As of May of 2014, the International Telecommunications Union (ITCU) estimates there are 6.9 billion mobile subscriptions worldwide representing approximately 95.5% of the world population.[2] Of these subscriptions, the ITCU estimates 5.4 billion of them are in the developing world, and that the underlying total 6.9 billion subscriptions are approximately 4.5 billion mobile users (or 62% of the world's population).

Total estimated worldwide sales for all handsets in 2014 are approximately 1.845 billion, growing to approximately 1.89 billion in 2015.[4] Sales of smartphones are estimated at 1.24 billion units in 2014,[3] are expected to grow to 1.5 billion in 2015,[4] and to 1.8 billion by 2018, representing a CAGR of 12.3% for the period 2013–2018.[3] Whereas growth of the overall handset market has slowed to rates consistent with a replacement model, sales of the smartphone segment continue to grow at a healthy rate.

History has shown that the set of top handset OEMs in terms of market share has seen considerable churn over time. Recent consolidations in the industry with Microsoft acquiring Nokia's handset business and Lenovo acquiring Motorola's handset business have resulted in a continuation of this trend. For the first time in 2015, Nokia (Microsoft) is not expected to be in the top five handset OEMs for the year.[4] As China has continued to expand its mobile communications infrastructure and major Chinese carriers roll out 4G networks, Chinese handset OEMs have moved into leadership positions in terms of market share. According to Canaccord/ Genuity,[4] the top five OEMs in 2015 in terms of market share for all handsets sold are expected to be (in order): Samsung, Apple, Huawei, Lenovo, and Xiaomi. The same vendors will make up the top five in smartphone sales with Lenovo and Xiaomi trading places for the number four and five positions.

With this overview of the mobile handset market, let us now focus in more closely on the history and status of the RFFE architecture of the mobile handset.

1.3 EVOLUTION OF HANDSET RF FRONT-END ARCHITECTURES

Broadly speaking, we can define the RFFE of a mobile handset as the collection of hardware components located between the leading edge of the transceiver and the various antennas in the phone. While this real estate typically also includes the front-end components for Wi-Fi connectivity and GPS, we will focus solely on the cellular RFFE and include the associated antennas as a part of the discussion (since, as we will see later, tunable RF solutions both enable and potentially require, new approaches to antenna design and new levels of antenna functionality). From an architectural perspective, the RFFE can be viewed as one or more chains of RF devices whose purpose is to enable and optimize high-frequency transmission and reception of voice, data, and control signals (connectivity) between the handset and the network. The functions performed by each RF chain include power amplification, selection and rejection of signals (filtering), the actual radiation and reception of signals by the antenna, and the switching of signals.

While the transceiver, baseband, applications processor and other digital and relatively lower frequency components in the mobile handset have benefited greatly from continuing advances in complimentary metal oxide semiconductor (CMOS) technology and Moore's law scaling, components in the RFFE have only recently and selectively begun to enjoy similar benefits. Why has this been the case? Potential reasons include the following:

- RF components must meet very demanding performance requirements in order to ensure the highest quality, most reliable connection to the network possible. To do so, the various RFFE functions have traditionally been designed to be frequency and function specific. Selection of an implementation approach has also been driven by performance requirements leading to a heterogeneous collection of technologies in the RFFE including at various points in time acoustics, ceramics, magnetics, and gallium arsenide semiconductors. It has only been in recent years that CMOS technology has advanced, in combination with circuit and architecture design techniques, to the point that selective usage in the RFFE has become viable.
- From the consumer's point of view, the RFFE is an invisible portion of the mobile handset's architecture. As the Apple iPhone "Antennagate" episode demonstrated, the average consumer is not aware of or even generally interested in "how" their phone actually works, but cares deeply about attributes that affect their end-user experience—battery life, dropped calls, roaming capability, data speeds and quotas, and cost. Ironically, these attributes are either determined or are greatly affected by the performance and capabilities of the RFFE of the handset.

This is not to say that there have not been consistent advances in the RFFE architecture. There have been albeit at a much slower pace than in the digital portions of the handset. Newer combinations of design techniques and technologies have largely supplanted some of the widely used specialty technologies in the past such as magnetics. Other technologies have continued to scale in size while maintaining (or improving) performance resulting in physically smaller, more power efficient, and lower cost solutions. And advances in passive device, substrate, and packaging technologies have fueled continued integration at the module level providing multiband and multimode capabilities within a single footprint. But a new set of requirements and constraints in the form of 4G or LTE and LTE-Advanced (LTE-A) technology

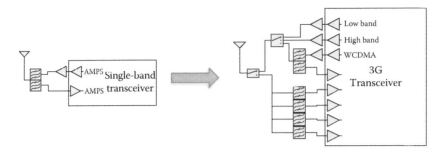

FIGURE 1.3   1G versus 3G RF front-end complexity.

has clearly highlighted that a new approach to scaling and implementing the RFFE is required.

From the initial deployments of first-generation cellular technology, through today's explosive growth of 4G networks, one item has remained consistent—find ways to do more at the product level with less (space, volume, power, cost, time). Without general access to the benefits of CMOS technology, scaling the RFFE has been achieved historically by duplication. Need support for a new frequency band or mode of operation? Add an additional chain of RF components targeted at the specific frequency or mode. Repeat as required. Figure 1.3 provides a visual comparison between the structural complexity of typical "1G" and 3G RFFEs.

While some of the advances in RFFE implementation mentioned earlier allowed this approach to remain the de facto scaling technique up until the 3.5G time frame, new form factor, performance, connectivity, power, time-to-market, and loss considerations began to mitigate the benefits and limit usefulness with 4G.

Table 1.1 shows a snapshot of the frequency band plan for 4G/LTE on a global basis. Once this plan is fully implemented, a truly global handset could need to provide support for up to 43 different frequency bands, up from an average of approximately 10 bands in 2010.[6] In addition to the large increase in band count, 4G targets maximizing network capacity by increasing peak data rates to the 100 Mbps to 1 Gbps range depending on the mobility of the user. To support these data transfer rates, LTE-A includes support for multiple input/multiple output (MIMO) architectures employing multiple antennas and carrier aggregation (CA). CA is a new capability which allows (depending on the LTE release level) between two and five channels to be stitched together dynamically in software to provide a larger data "pipe" for reception, transmission, or both. While enabling tremendous potential to maximize network capacity, CA brings significant challenges to the RFFE, which has relied heavily on fixed channel bandwidths, known spacing between transmit and receive frequencies, and single feed antennas. The number of CA combinations being planned and deployed is growing rapidly, and differs between geographies and carriers making transparent global roaming a harder goal to achieve. CA can be implemented using multiple channels within a band ("intra-band" CA) or using different channels in different bands ("inter-band" CA), which along with transmit CA, sets new, tougher requirements for the linearity and harmonic performance of the RFFE in 4G handsets.

TABLE 1.1

LTE Frequency Bands and the Corresponding Regions[5]

LTE Bands	Uplink (MHz)	Downlink (MHz)	Duplex Spacing (MHz)	BW (MHz)	Duplex Mode	Deployment in the World
Band 1	1920–1980	2110–2170	190	60	FDD	China, Japan, EU, Asia, Australia
Band 2	1850–1910	1930–1990	80	60	FDD	North and South America
Band 3	1710–1785	1805–1880	95	75	FDD	EU, China, Asia, Australia, Africa
Band 4	1710–1755	2110–2155	400	45	FDD	North and South America
Band 5	824–849	869–894	45	25	FDD	North and South America, Australia, Asia, Africa
Band 6	830–840	875–885	45	10	FDD	Japan
Band 7	2500–2570	2620–2690	120	70	FDD	EU, South America, Asia, Africa, Australia
Band 8	880–915	925–960	45	35	FDD	EU, South America, Asia, Africa, Australia
Band 9	1749.9–1784.9	1844.9–1879.9	95	35	FDD	Japan
Band 10	1710–1770	2110–2170	400	60	FDD	North and South America
Band 11	1427.9–1447.9	1475.9–1495.9	48	35	FDD	Japan
Band 12	698–716	728–746	30	18	FDD	North America
Band 13	777–787	746–756	31	10	FDD	North America
Band 14	788–798	758–768	30	10	FDD	North America
Band 17	704–716	734–746	30	12	FDD	North America
Band 18	815–830	860–875	45	15	FDD	North and South America, Australia, Asia, Africa
Band 19	830–845	875–890	45	15	FDD	North and South America, Australia, Asia, Africa
Band 20	832–862	791–821	41	30	FDD	EU
Band 21	1447.9–1462.9	1495.9–1510.9	48	15	FDD	Japan
Band 22	3410–3500	3510–3600	100	90	FDD	
Band 24	1626.5–1660.5	1525–1559	101.5	34	FDD	

(*Continued*)

TABLE 1.1 (*Continued*)
LTE Frequency Bands and the Corresponding Regions[5]

LTE Bands	Uplink (MHz)	Downlink (MHz)	Duplex Spacing (MHz)	BW (MHz)	Duplex Mode	Deployment in the World
Band 33		1900–1920	NA	20	TDD	
Band 34		2010–2025	NA	15	TDD	China
Band 35		1850–1910	NA	60	TDD	
Band 36		1930–1990	NA	60	TDD	
Band 37		1910–1930	NA	20	TDD	
Band 38		2570–2620	NA	50	TDD	EU
Band 39		1880–1920	NA	40	TDD	China
Band 40		2300–2400	NA	100	TDD	China, Asia
Band 41		2496–2690	NA	194	TDD	
Band 42		3400–3600	NA	200	TDD	
Band 43		3600–3800	NA	200	TDD	

Figure 1.4 provides a schematic for a 4G RFFE implementation that supports 2×2 MIMO and 20 frequency bands. The structural complexity of this implementation provides clear evidence as to why the traditional linear, chain-by-chain approach of scaling the functionality of the RFFE is not viable moving forward.

But, what if we could scale in a superlinear fashion providing the required performance and functionality, while realizing savings in power, space, loss, time-to-market, and cost, along with unprecedented flexibility?

1.4   WHAT IS TUNABLE RF?

Although discrete tunable RF functions and products have existed for many years, a new generation of such devices and circuits have been pioneered and applied to the RFFE design of mobile handsets by the authors of this book over the past decade. As used in this context, we define a tunable RF function to be one whose target frequency of operation and/or performance parameters are dynamically adjustable under software control. Such adjustments are typically in response to changes in the micro- and/or macro-environment of the handset. Representative changes include but are not limited to, changes in: the physical configuration of the handset; hand and/or head placement; transmit power and/or receive sensitivity; the local interference environment; and call hand-offs or other types of channel/band/mode reassignments by the network. On occasion, the types of tunable RF functions we will discuss here are referred to as programmable RF, or digitally tunable or digitally programmable RF. We will treat all these terms as synonyms and refer to them collectively as tunable RF.

We can further characterize tunable RF functions for mobile handsets by two additional parameters: (1) the amount of available tuning capability, and (2) the method of control. Tuning capability can be described in terms of precision, number of steps, and range of coverage while control can be broadly viewed as either open

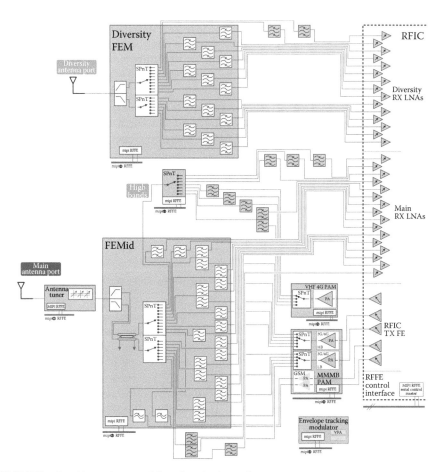

FIGURE 1.4 4G RF front-end functional schematic.

loop or closed loop. As an example, we can look at a tunable RF filter function. For such a filter, we can think of the range of coverage as being the range of frequency bands the filter can be tuned to cover. The number of steps would denote the number of discrete filter settings we can realize between a minimum and a maximum value (the range of coverage) while the precision would indicate how precisely we could set each filter value or state (± a percentage value). If the filter were being tuned in operation by the baseband processor (a typical scenario) using a predefined lookup table, we would refer to the filter as being controlled in an open-loop fashion. Alternatively, if real-time information from the environment is being used to control the filter (either directly or via the baseband), we would call this closed-loop operation. As with most options, decisions between open- and closed-loop control are based on many variables including complexity, ease of integration, and cost.

To demonstrate another compelling advantage of tunable RF, we can expand on our tunable RF filter example. Let us suppose that a certain handset needs to support the following LTE (low) frequency bands: (1) Band 17 (704–746 MHz);

(2) Band 13 (746–787 MHz); and (3) Band 5 (824–894 MHz). Using traditional fixed-frequency filters, we would need to implement one filter (in this case, one duplexer) per frequency band whereas if we had a tunable RF filter supporting the required levels of performance and which could be tuned between 704 and 894 MHz with a sufficient number of steps and precision, we could replace the three fixed filters with one tunable device. Now let us further suppose that the handset supplier desires to supply the same model of phone in a different geography or to a different carrier who desires to cover Band 20 (791–862 MHz) instead of Band 5. The same tunable RF filter can be used to cover the B17/B13/B20 combination in the second phone as the B17/B13/B5 combination in the first phone. All that is required is a change in software and/or lookup table contents. Thus, tunable RF provides the potential for the type of superlinear scaling the RFFE of handsets needs to meet the challenges of multi-geography (or super-geography) and global LTE handsets by enabling one physical tunable function to replace multiple instances of its static counterpart. Correspondingly, as shown in Figure 1.5, one tunable RF chain can replace multiple fixed-frequency or mode-specific chains of components. In the limit (assuming sufficient tuning capabilities), the number of physical chains of RF components can be reduced to the maximum number of radio paths that can be active concurrently.

Tunable RF has the capability to assist the designer in meeting an increasingly difficult set of RFFE design constraints. Benefits to marketing can include realization of desired form factors (industrial designs) including increasingly thinner designs, all glass/screen fronts, and metal case/frame construction. Equally, if not more so, are the business benefits available to the handset OEM. These include faster time to market, superior performance and usability, reduction in the number of SKUs, reduction in bill-of-materials (BOM) cost and complexity per handset, and the possibility of field upgrades to accommodate access to a new or re-farmed spectrum.

With all of the potential advantages provided by tunable RF, why have all handset OEMs not adopted a tunable approach to implementing the RF front end? Reasons include the following:

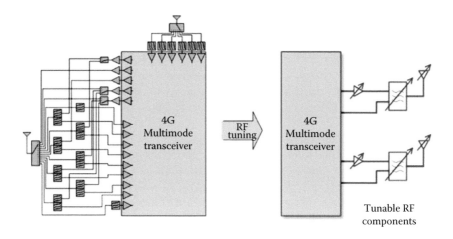

FIGURE 1.5   Traditional 4G RF front end versus tunable 4G RF front end.

- *Risk*: A delay in time-to-market or a problem in the field can have huge negative and long-lasting, financial consequences. Everyone is happy to be a fast follower as long as someone else is the pioneer when the risk is sufficiently high.
- *Different approach*: To maximize the potential of a tunable RF solution, the front end needs to be designed to be tunable which generally implies a different set of design criteria and goals than with traditional fixed RF components. For example, designing an antenna to be tunable means focusing on maximizing radiated efficiency and allowing the tuner to do the matching as opposed to having to balance radiation efficiency and matching as is traditionally done.
- *Cost*: Handset OEMs have large investments in existing hardware platforms that they need to leverage. And with severe cost pressure from their network operator customers, they are driven to avoid any possibility of BOM cost increase unless there is a compelling reason for adoption.

Compelling reasons for adoption can come in many forms. Whereas there were clear benefits to be realized with the adoption of tunable RF solutions for 3.5G and 3.9G phones, it was not clear that adoption was more than a nice-to-have feature. But with 4G, there is a clear mandate for implementation of tunable solutions. Lower operating frequencies, higher data rate expectations from consumers enabled by CA and MIMO architectures, return on investment (ROI) expectations from network operators on investments in 4G infrastructure, and continued ID design constraints have resulted in a clear call for the evolution of the RF front end.

However, given the handset OEM concerns over risk, differences in implementation approach, and cost—How have tunable RF suppliers driven the adoption of tunable components? By identifying an initial application where there is a high need and substantial benefit which still allows the OEM to maintain their platform investment—antenna tuning. Rather than starting in the middle of an existing RFFE architecture (where a fan-in/fan-out challenge is created by inserting a tunable component between static components), the antenna is located on one end of the RFFE. Figure 1.6 shows a WiSpry impedance tuner located on a production phone board next to the antenna contacts (and at one end of the handset).

Given the location, the addition of a tuner is possible with minimum perturbations to the remainder of the existing RFFE design. This lowers the barrier to adoption and as we will see in the next section, has driven initial production revenue growth of tunable RF.

Antenna tuning can be categorized as follows based on the physical location/connection of the tuner and the intended function:

- *Aperture or load tuning*: The tuner is directly connected to the radiating element at some location physically distant from the feedpoint.
- *Impedance tuning*: The tuner is in-line with the transmission line to the radiator feed; signal lines are shielded (so as to be nonradiating) at the tuner location.
- *Feedpoint or coupling tuning*: The tuner is located in the transition between the transmission line to the radiator feed and the radiating element.

FIGURE 1.6   Placement of an impedance tuner in an RF front end.

We can also categorize antenna tuners based on the technology used for their implementation. Today, three basic technology approaches are being used for the implementation of antenna tuning solutions[7]:

- *Switched capacitors*: Typically, SOI FETs are used to switch between fixed MIM capacitors.
- *Barium strontium titanate (BST) capacitors*: Are used to provide a smoothly varying change in capacitance through a combination of a control chip and one or more BST capacitor die.
- *RF microelectromechanical systems (MEMS) capacitors*: These use mechanical structures that are movable (actuated) using electrostatic forces to provide a set of digitally tunable capacitor values, typically monolithically integrated in a CMOS process.

Figure 1.7 provides an example of the implementation of an array of RF-MEMS capacitors in a CMOS process. In Figure 1.7a, the capacitor in the rear is shown in its minimum, unactuated, capacitance state while the capacitor in the front has been deflected vertically toward the substrate using a DC voltage to provide the maximum capacitance value. Figure 1.7b shows an array of such devices in which each individual capacitor is addressable providing for multiple tuning states (or capacitance values) from the minimum to the maximum value of capacitance for the entire array.

The advantages and disadvantages of each of the three implementation approaches currently used for antenna tuning are described in detail in Steel and Morris.[7] Figure 1.8 provides a graphical summary derived from their comparison of the implementation technologies.

The various categories of antenna tuning and the implementation technologies in use today will be explored in depth in later chapters in this book, along with numerous other types of tunable functions, systems, and concepts.

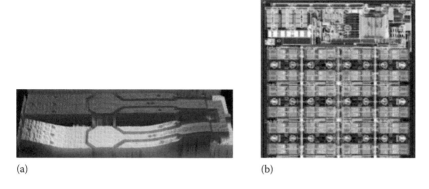

FIGURE 1.7 WiSpry RF-MEMS capacitor pair (a) and (b) array of 48 capacitors with an integrated charge pump and digital control logic implemented in an 180 nm RF-CMOS.

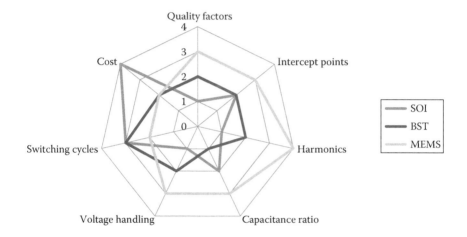

FIGURE 1.8 Comparison of tunable RF implementation technologies.

1.5 MARKET SNAPSHOT: RF FRONT-END AND TUNABLE RF

The RFFE total available market (TAM) for mobile handsets is a very large market. While various sources disagree on the exact composition of the TAM, current size, and growth rate, they all agree that the market is growing at a rapid pace.[8] Estimates of the current TAM are in the $9 billion range in 2014–2015.[9,10] Some sources[6,9] show the TAM growing to approximately the $13 billion range in 2017–2018, while more aggressive forecasts,[10] show 20% year-on-year growth forecasts for 2015–2016 leading to a total 2016 TAM of approximately $14.5 billion. This growth is being fueled by the continued deployment and growth of 4G networks worldwide. Within the RF front-end TAM, the largest category of components is filters and various forecasts[9,10] show that filters will make up somewhere between 50% and 60% of the total RF front-end TAM during the 2016–2018 time frame. As with the overall RFFE market, LTE filters, in particular duplexers, are driving the growth of the filter category.

It is also instructive to look at estimates of the total RF front-end BOM cost for various generations of handsets. Estimates for total 2G BOM costs are in the $0.50–$0.55 range.[6,10] 3G–3.9G RF front-end BOM costs range from $3.75 to $4.00.[10,11] Depending on whether a 4G handset is categorized as basic or regional, or equipped for global roaming, recent RF front-end BOM estimates fall in the ranges of $7.00–$8.00, approximately $10.50, and from $13.25 to as high as $17.00 respectively.[6,10–12] And a recent estimate of the weighted average RF front-end BOM cost across the different SKUs of the iPhone 6 comes in at $14.65 per handset.[12] Given that RF component and module costs have typically fallen year-on-year, the approximate 28× increase in total BOM cost between a 2G handset and the average iPhone, and the approximate 3× increase between a 3G handset and the average iPhone, are strong indicators of the explosion in complexity and component count in the front end, along with the rapid increase in performance requirements for these components.

As mentioned in Section 1.4, the adoption of tunable RF in mobile handsets today is being led by antenna tuning applications. As tunable RF is a new and emerging segment of the overall RFFE market, present analyst coverage is limited. Among the earliest published market forecasts is a projection from Oppenheimer in early 2013,[13] which showed a 2015 TAM for antenna tuners of $772 million growing to slightly over $1 billion by 2016. A more recent report by Mobile Experts[9] projects 2015 revenues of $112 million growing to $270 million by 2018 as shown in Figure 1.9.

Strategic Analytics[14] predicts sales of approximately 750 million units in 2015 growing to 1.9 billion units by 2018. Meanwhile, Yole Development's estimates of the penetration of MEMS tuners in the market are for a total of 16 million units in 2015 growing to approximately 77 million units in 2017.[15]

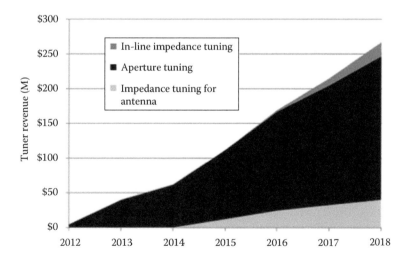

FIGURE 1.9 Total revenue projection for discrete tunable components in the total RF front-end market.

Finally, according to Taylor,[14] the estimated market size for power amplifier (PA) envelope tracking in 2015 is 251 M units increasing to 1 billion units in 2018. And Yole[15] forecasts initial sales of MEMS-based tunable filters and PAs starting in 2017.

Based on the market data presented earlier it is clear that adoption of tunable RF, while growing rapidly, is still very much at an early stage. Meanwhile, the overall RF front end TAM continues to grow at a substantial rate from its approximate $9 billion base creating both a tremendous opportunity and need for rapid innovation and evolution. Figure 1.10 provides a graphical summary of some of the major innovations in handset architecture over the 14-year period between 1998 and 2012.[14] As impressive as these achievements are, more progress is needed at a faster rate to keep pace with the demands of 4G users and networks. In the next section, we explore how we can best accelerate the evolution of the RFFE through the adoption of tunable RF.

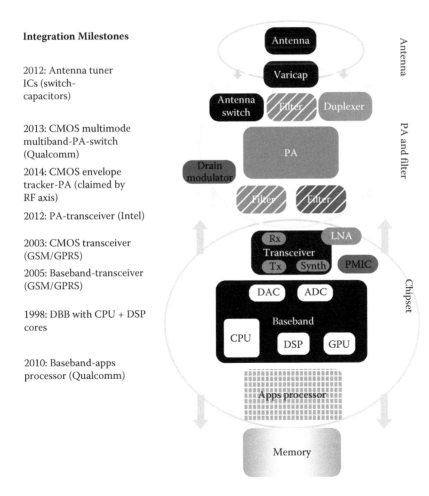

FIGURE 1.10 Mobile handset integration milestones.

1.6 TUNABLE RF COMPONENT AND CIRCUIT OUTLOOK

Now that tunable RF has established a beachhead in the mobile handset front end in the form of antenna tuners, the next steps in market penetration need to be identified and taken. We could turn toward the antenna and look to integrate antenna tuning and antenna design into an overall solution. And this is exactly what antenna firms such as Skycross and Ethertronics have done as they have gone to market with active and adaptive antenna solutions. Or, we could turn toward the filter side of the front end to see how and if we might take advantage of the existing multibillion dollar opportunity that today is pretty much the exclusive providence of acoustic devices. Or perhaps, as a third alternative, we should next address the potential opportunity to provide tunability in the power amplification segment. If it were not for the performance requirements of filters, the choice would perhaps be obvious. But antennas and PAs/LNAs both have their own sets of technical and business challenges and neither provide the upside opportunity of the filter segment.

By choosing to focus on filters we face another challenge; none of the existing implementation technologies for tunable RF—MEMS, BST, or SOI—enable sufficient broadband, brick wall filtering on par with SAW/BAW/FBAR filters. And of the three, only series-configured MEMS devices appear to provide performance levels that, when combined with architectural innovation or utilized in more restrictive applications, can provide the required filtering in the RFFE.

To begin to explore the opportunity in the RF filter segment of the market, WiSpry has developed prototype tunable RF notch filters for selected coexistence applications. These filters provide an additional 20–30 dB of rejection on demand for situations where interfering signals are well defined (for example, Wi-Fi and high-band cellular transmitters) and relatively closeby. Utilizing the wide tuning range of MEMS tunable capacitors, these filters can be tuned far out of band, in effect turned off, to avoid the insertion loss of fixed filters incurred during the majority of time when there are no coexistence issues to be handled. When "turned on," losses are comparable to those of static filters.

Expanding on the concept of tunable, narrowband rejection, Figure 1.11 shows initial measured performance results from a prototype dual multiband tunable (TX or RX) filter module developed by WiSpry.

An examination of the results shows that the notch response provides a reasonable level of tunable rejection across multiple bands.

Continuing this bottom-up approach, development of a prototype tunable narrowband front-end architecture has been underway for several years in a consortium effort between WiSpry, Intel, and Aalborg University in Denmark. Figure 1.12 shows an initial hardware demonstrator of a complete multiband tunable RFFE architecture covering Band 1 through Band 7. The demonstrator contains a pair of TX and RX antennas which, when combined with the tunable TX and RX notch filters, results in a tunable duplexing capability. This prototype, now several years old, achieved most of the 3GPP specifications using only eight tunable hardware components (including the antenna pair), and without the use of any acoustics or switches. The demonstrator board, which is pretty much empty space, has the dimensions of a standard mobile handset phone board. Development of this architecture has continued and subsequent prototypes covering both low band and high band are now operational in the lab.

Figure 1.13 provides a schematic representation of a complete LTE-A capable tunable RFFE concept that WiSpry refers to as CAFE (Compact, Agile Front End). As shown, this architecture inherently supports LTE CA and MIMO functions,

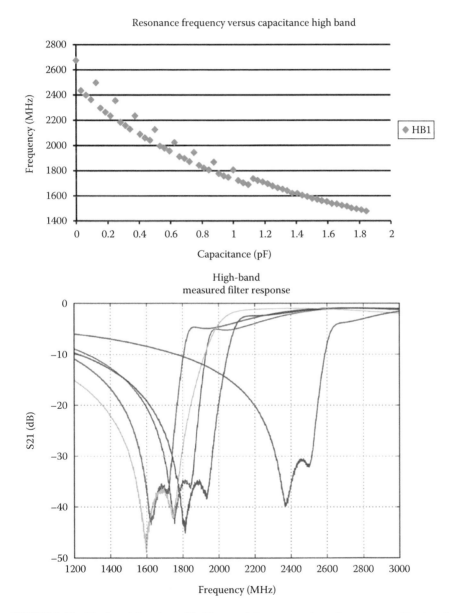

FIGURE 1.11   Dual multiband tunable filter module measured results.   (*Continued*)

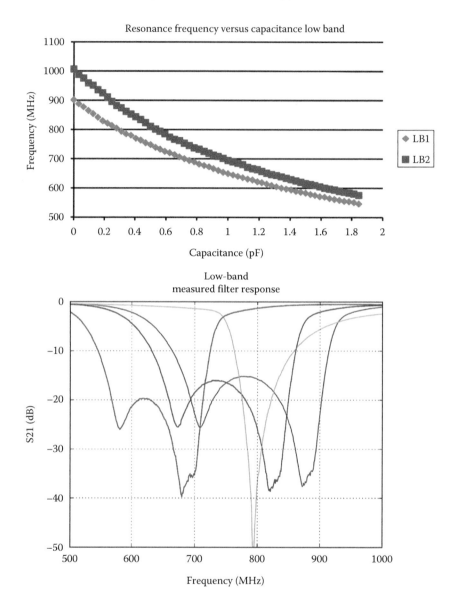

FIGURE 1.11 (***Continued***)   Dual multiband tunable filter module measured results.

requiring only one additional tunable RF path for each potentially concurrently active radio link.

Viewed in the context of the $7.00–$17.00 BOM opportunity for RF front ends in 4G handsets, architectures such as CAFE look compelling. But at the same time, they are revolutionary and, as such, they carry a great deal of perceived, and some actual, risk. Fundamentally, one of the requirements of such an architecture is for production, radiated test of the handset. Great progress toward this capability is

FIGURE 1.12   Prototype multiband tunable RF front-end architecture.

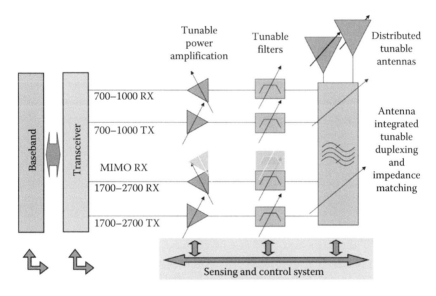

FIGURE 1.13   Fully tunable MMMB RF front-end architecture model.

being made by companies such as LitePoint (see Chapter 14), but it will take several years until the necessary certification test cases and specifications are ratified, and the production test infrastructure is in place. In the interim, tunable RF can continue to grow market share and revenue bottom up through the implementation of additional and next-generation antenna tuners, tunable antennas, tunable PAs and LNAs, tunable filters for applications such as coexistence, and perhaps subsets of CAFE-like architectures on the receive side, while pursuing the top-down development of a complete tunable narrowband front-end solution in parallel.

Ultimately, it is likely that no single design approach or technology will prove to be the silver bullet but rather that future RFFE architectures will evolve to encompass the best of both fixed and tunable functions, along with CMOS, acoustic, and other existing and new implementation technologies. Whatever form progress takes, time is of the essence. The year 2020 will be upon us sooner than we think, and with it, an entirely new level of even more daunting performance challenges with 5G.

1.7 CONCLUSIONS

Over the 42-year period since the first mobile phone call was placed, mobile communications has changed the world in ways that we could not have imagined. Mobile communications technology has evolved from voice communications over an analog link, to a digital, data-centric architecture that provides for high-speed downlink and uplink of data, downlink of high-speed streaming video, and voice over data. Communications standards have evolved to the present fourth-generation of technology, whose deployment in the market has ramped faster than any previous such standard. Today, an astounding 62% of the global population is estimated to be mobile phone users. Mobile handsets, and the RF semiconductors they employ have become the growth engine of consumer electronics and in turn, of the overall semiconductor industry. In 2015, it is estimated that a total of 1.8 billion new handsets will be sold, an increasing majority of which are smartphones whose processing, memory, and software capabilities will far surpass those of the leading PCs of just a few years ago.

As the mobile handset market has matured, different portions of the handset architecture have evolved at different rates. Whereas the lower frequency, digital portions of the handset have enjoyed the scaling benefits of Moore's law, the RF front end, the increasingly crucial link between the handset and the network, has evolved much more slowly. Increasingly more challenging performance specifications coupled with the rapid proliferation of frequency bands on a global basis with LTE, have positioned the RFFE as a key bottleneck to the next stages of innovation in mobile communications.

To address this bottleneck, a new design paradigm has emerged over the past decade. The concept of tunable RF, high performance RF components whose frequency and functions can be dynamically adjusted via software, has entered the front-end architecture initially in the form of antenna tuners. Such tunable RF components enable the replacement of multiple components with a single hardware device and reduce the number of chains of such components required to support multiband operations to a number equal to that of the maximum number of concurrently active radios. From this perspective, we can consider tunable RF front ends as the final step in the implementation of software defined radios.

The current tunable RF market is in the early stages of growth with only a limited number of players in production revenue. Total revenues for the segment are measured in the hundreds of millions of dollars, a small percentage of the overall $9 billion RFFE market, which is growing rapidly. Of the three implementation technologies being utilized for tunable RF devices today—MEMS, BST, and SOI—the majority of the current market belongs to antenna tuners implemented using SOI FET switches.

Looking ahead, tunable RF can be expected to grow faster than the RFFE market. Antenna tuners will enjoy deeper penetration with average utilization per handset pushing well beyond one per unit; antenna vendors will bring new types of active and adaptive antenna systems to market that meet the volume constraints imposed by desirable industrial handset designs while providing the levels of over-the-air performance required to keep both consumers and network operators satisfied. MEMS-enabled tunable filters will roll out in the market and advances in the application of tunability to power and low-noise amplifier applications will become apparent. Finally, and perhaps most dramatically, completely new fully tunable RFFE architectures will begin to appear on the market, most likely beginning with coexistence filtering and diversity receive applications. Once the required standards and production manufacturing capabilities supporting radiated test are deployed, these architectures will complete their evolution into software defined, narrowband designs, with broadband tunability. Such architectures will most likely be implemented using hybrid approaches of fixed and tunable devices implemented in CMOS and specialty technologies as most appropriate.

And finally, it is estimated that in the 2020 time frame, commercialization of 5G technology will begin, bringing a new wave of challenges and opportunities as wireless communications and connectivity continue their seemingly inevitable march toward becoming so pervasive and tightly integrated in the fabric of society as to become invisible in daily life.

REFERENCES

1. Young, J., Mobile front end module—The battle between SIP & SOC, *The 12th International System-on-a-Chip (SOC) Conference, Exhibit, and Workshops*, Irvine, CA, October 2014.
2. International Telecommunications Union, Measuring the Information Society Report, May 5, 2014, pp. 2–3.
3. IDC Press Release, Smartphone momentum still evident with shipments expected to reach 1.2 billion in 2014 and growing 23.1% over 2013, May 28, 2014, www.idc.com.
4. Canaccord/Genuity, Smartphone demand, *Daily Letter*, pp. 2–3, November 3, 2014.
5. Derived from https//en.wikipedia.org/wiki/E-UTRA, Frequency bands and channel bandwidths, accessed April, 2015.
6. Qorvo 2014 Investor Day Presentation, www.qorvo.com, accessed April 2015.
7. Steel, V. and Morris, A., Tunable RF technology overview, *Microwave Journal*, 55(11), 26–36, November 2012.
8. BMO Capital Markets, Digging into China, handset model update, Communications Equipment Report, May 8, 2014.
9. Madden, J., Private communication, March 2015.
10. Charter Equity Research, Post December 2014 quarter wireless and semiconductor industry analysis, February 13, 2015.
11. Barclays, Triquint semiconductor, filters still the best way to play LTE, Equity Research, November 1, 2013.
12. Charter Equity Research, iPhone 6: A detailed teardown analysis, September 19, 2014.
13. Oppenheimer, LTEvolution, The implication for wireless semiconductor vendors beyond LTE, Equity Research, March 25, 2013.
14. Taylor, C., Radio front-end integration driving cellular component market dynamics, *Strategic Analytics Presentation*, Private communication, October 22, 2014.
15. Yole Development, RF MEMS market 2015 update, Private communication, March 2015.

2 RFSOI Technologies on HR Silicon Substrates for Reconfigurable Wireless Solutions

Julio Costa

CONTENTS

2.1 INTRODUCTION TO RFSOI ON HIGH-RESISTIVITY SILICON TECHNOLOGY

Tunable and reconfigurable applications using RFSOI-on-HR (high-resistivity) silicon technologies are being deployed today in the radio frequency (RF) section of virtually every smartphone platform driven by the increasing data rates demanded by the 4G RF cellular front ends. RF cellular systems are becoming increasingly complex with numerous transmit and receive bands, the possibility of multiple antennas, and new architectures that involve uplink and downlink carrier aggregation. Such new architectures present extreme challenges for conventional fixed-band systems composed of PAs, switches, and filters, driving the need for reconfiguration and tunability. In a relatively short time since its introduction in the early 1990s, RFSOI-on-HR silicon has proven to be a cost-effective high-performance, and high-yielding technology offering the RF designer the needed versatility for multiple design RF device configurations. This chapter will present a chronology of RFSOI technology development in the wireless industry, as well as the basic device fabrication and relevant physics, the different approaches to switch cell design, and traditional trade-offs, including relevant RF device metrics such as linearity, insertion loss, and intermodulation distortion. This chapter will also cover typical configuration of tunable RFSOI switch cells in tunable networks, including novel uses of RFSOI capabilities in cellular 4G systems as well as the critical specifications needed in this application space.

The chapter will focus entirely on mainstream RFSOI technologies fabricated on high-resistivity (HR) silicon substrates, since these technologies are readily available from many high-volume silicon foundries, leaving out other captive offerings for wireless RF switching such as silicon-on-sapphire (SOS) or microelectromechanical systems (MEMS) that also offer equivalent tunable switching devices for the RF mobile space. RFSOI-on-HR silicon technologies are now offered through the very popular MOSIS (www.mosis.com) multi-project-wafer (MPW) shuttle service, which gives low-cost access to a high-performance RFSOI technology to virtually anyone in industry/government/academia. Although the majority of applications for RFSOI in the mobile space today consist of switching solutions (RF switches, impedance, and aperture tuners), several RFSOI-on-HR linear power amplifiers (PAs) are also starting to gain acceptance, suggesting high levels of integration in future RFSOI product roadmaps.

It is important to note that the maturity of RFSOI-on-HR silicon technologies in the mobile marketplace did not happen overnight. It was the product of constant technological advances and excellent collaboration between foundries, SOI substrate wafer suppliers, and the RF device community, showing very steady progress since the first attempts were made in the early 1990s. In this chapter, we discuss some of the many challenges RFSOI-on-HR silicon has gone through in the last two decades to solidify its dominance in the cellular RF front end.

2.1.1 TECHNOLOGY DRIVERS FOR RFSOI-ON-HR SILICON SUBSTRATES

Exponential consumer data demand in mobile applications is the key driver for growth in RFSOI-on-HR technologies. The chart in Figure 2.1 clearly depicts this trend: thousands of new "Apps" in the smart handset now demand fast download/upload of consumer data, to the point that we now speak in terms of "exabytes"/month, where one EB = 1×10^{18} bytes, essentially equivalent to one billion people in the world downloading 1 GB per month. Such high data rates require RF front-end components—those that directly interface with the antenna and the transceiver blocks—to operate at high power levels with a very high degree of linearity.

Mobile data, of course, must be delivered fast and accurately with a high quality of experience, and constantly new cellular standards continue to be implemented by the mobile industry to meet this increasing demand. Such requirements necessitate unparalleled cooperation between carriers, infrastructure equipment, and the handset manufacturers.

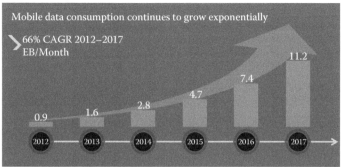

1 EB = 10^{18} bytes = 1 billion GB

FIGURE 2.1 Mobile data growth. (From RFMD Cellular Group; CISCO's public website, http://www.cisco.com/c/en/us/solutions/collateral/service-provider/visual-networking-index-vni/white_paper_c11-520862.html.)

Cellular handset and infrastructure providers have embraced next-generation 4G system architectures at an unparalled rate of adoption, faster than anything comparable in the history of this marketplace. At the time of this publication, fourth-generation (4G) mobile cellular standards have been deployed throughout most of the world, with the adoption of Long Term Evolution (LTE) modulation to achieve effective download speeds of 10–20 Mbps, comparable to most residential LAN network data rates; plans are being developed by the mobile industry for fifth-generation (5G) mobile standards with downlink speeds of hundreds of Mbps with the addition of several new transmit/receive bands as well as the incorporation of carrier aggregation (CA). The evolution of these mobile standards translates into ever more complex transmission waveforms at many different frequency bands— latest count around 45 bands worldwide—which create numerous challenges in the components that comprise the radio of the cellular handset, in particular, linear RF power, harmonic generation and intermodulation distortion products. Figure 2.2 illustrates the evolution of these cellular standard waveforms, with particular attention to the increasing peak-to-average power ratio (PAPR) demands of the newer generation of cellular standards.

FIGURE 2.2 Illustration of peak-to-average power ratios (PAPRs) for 3G, 3.5G, and 4G (LTE) waveforms.

The increase of PAPR values impose severe requirements in the voltage and power handling as well as harmonic generation and reliability of the technologies used in the front end of mobile systems.

Figure 2.3 lists the current frequency band spectrum allocation to the different bands, as defined by the 3GPP Mobile Broadband Standard Group.*†

This group is a consortium of industry, academia, and government agencies that is responsible for the definition and standardization of the communication protocols of the different bands used in mobile applications. The characteristics and definition of each frequency band is an intricate product of several different factors, owing to the fact that spectrum licenses are allocated in each separate countries to different carriers and operators.

The introduction of carrier aggregation (CA) in the newer 4G mobile releases added a significant layer of complexity and enhanced linearity requirements on all RF front-end components. In the first releases of carrier aggregation, a second band

LTE TDD frequency bands

Band	Uplink MHz		Downlink MHz		Width	Duplex	Gap
1	1920	1980	2110	2170	60	190	130
2	1850	1910	1930	1990	60	80	20
3	1710	1785	1805	1880	75	95	20
4	1710	1755	2110	2155	45	400	355
5	824	849	869	894	25	45	20
6	830	840	865	875–	10	35	25
7	2500	2570	2620	2690	70	120	50
8	880	915	925	960	35	45	10
9	1749.9	1784.9	1844.9	1879.9	35	95	60
10	1710	1770	2110	2170	60	400	340
11	1427.9	1447.9	1475.9	1495.9	20	48	28
12	698	716	728	746	18	30	12
13	777	787	746	756	10	-31	21
14	788	798	758	768	10	-30	20
15*	1900	1920	2600	2620	20	700	680
16*	2010	2025	2585	2600	15	575	560
17	704	716	734	746	12	30	18
18	815	830	860	875	15	45	30
19	830	845	875	890	15	45	30
20	832	862	791	821	30	-41	11
21	1447.9	1462.9	1495.9	1510.9	15	48	33
22	3410	3490	3510	3590	80	100	20
23	2000	2020	2180	2200	20	180	160
24	1626.5	1660.5	1525	1559	34	-101.5	67.5
25	1850	1915	1930	1995	65	80	15
26	814	849	859	894	35	45	10
27	807	824	852	869	17	45	28
28	703	748	758	803	45	55	10

LTE DD frequency bands

Band	Uplink MHz		Downlink MHz		Width
33	1900	1920	1900	1920	20
34	2010	2025	2010	2025	15
35	1850	1910	1850	1910	60
36	1930	1990	1930	1990	60
37	1910	1930	1910	1930	20
38	2570	2620	2570	2620	50
39	1880	1920	1880	1920	40
40	2300	2400	2300	2400	100
41	2496	2690	2496	2690	194
42	3400	3600	3400	3600	200
43	3600	3800	3600	3800	200
44	703	803	703	803	100

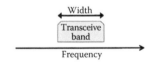

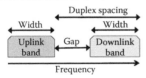

FIGURE 2.3   LTE frequency division duplexing (FDD) and time division duplexing (TDD) as defined in the 3GPP Mobile Standards Group in September, 2012.

* Official releases of the 3GPP Mobile Standards Group are kept at www.3gpp.org/releases.
† GPP Standards Update, Moray Rumney, http://www.home.agilent.com/upload/cmc_upload/All/25Oct12LTE.pdf?&cc=DE&lc=ger.

can be utilized at the same time to improve the downlink capacity of the handset to receive more data; TX CA will also be available in the near future to improve uplink capacity although currently a clear priority exists in the launch of RX carrier aggregation RF front ends since that is currently where the peak demand for data is concentrated. With CA, the technology requirements for RF devices become even more accentuated, especially for frequency division duplexing (FDD) band combinations where TX and RX functions are done at the same time. Figure 2.4 depicts the bands combinations that are planned for LTE 4G carrier aggregation.[1]

The inset of Figure 2.4 also illustrates why linearity requirements are much more demanding in the CA mode in the FDD operation; for this particular example of Bands 4/17 carrier aggregation—and there are numerous other troubling combinations—a harmonic caused by the TX B17 signal in the 700 MHz band can fall exactly on the RX B4 2100 MHz band. If this harmonic power level is sufficiently high, it will "jam or block" the intended RX signal. This unwanted harmonic product will, of course, look just like RX noise and the transceiver will be unable to receive the desired RX B4 signal. This particular CA problem is referred to as "carrier desense" and will invariably cause a condition where the communication between the mobile terminal and the base station is seriously affected.

Especially in FDD bands—where the TX/RX (transmit/receive) functions take place at the same time—there are extremely demanding requirements in the linearity of the devices and components that are on the TX/RX signal paths. Such concurrent RX/TX paths occur, of course, at the antenna of the handset and at all of the components in the downstream RF path directly connected to the antenna. The critical devices where such high linearity comes at an absolute premium are at

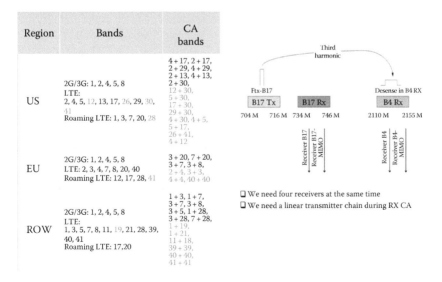

FIGURE 2.4 Carrier aggregation (CA) modes as defined in current 3GPP standards. The right side of the figure illustrates a potential problem created by having harmonics of the TX signal fall exactly on the band of the CA-RX signal.

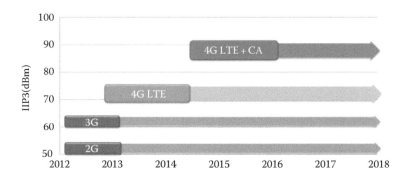

FIGURE 2.5   Required third-order input intercept point (IIP3) for RF front-end components as a function of evolving cellular standards. (IIP3: From Larry Schumacher, Intel Mobile Corporation, Challenges for radios due to carrier aggregation requirements, *International Wireless Industry Consortium* [*IWPC*] *Workshop*, November 6, 2012.)

the front-end (FE) switch and any of the impedance/antenna tuner components that are now commonly attached to smartphones being released in the cellular market. Transceiver chipset makers are anticipating that 4G LTE/CA modes will require RF front-end components with extremely high IIP3 (third-order input intercept point) values—in the 87–90 dBm range—substantially higher than any linearity level previously accomplished with any RF switching technology (see Figure 2.5). This will require that every possible source of distortion in the front-end components be thoroughly investigated and minimized to the absolute minimal degree. In addition to high linearity requirements, these front-end switching blocks are also required to have very low insertion loss for the ON state and high isolation for the OFF state.

Through intensive collaboration between academia, silicon wafer foundries, SOI substrate providers, and the mobile RF component makers, RFSOI-on-HR today has largely displaced traditional III–V and silicon-on-sapphire FET power switching solutions due to its improved performance, reduced size and excellent manufacturing yields, and low costs. RFSOI solutions are now present in essentially every major handset platform, providing state of the art performance in increasingly complex multi-throw RF switches, tunable RF capacitance arrays, and antenna RF tuners.

2.2   RFSOI-ON-HR SILICON SUBSTRATES AND DEVICES

The success of RFSOI-on-HR silicon in the RF front-end spaces owes a great deal to recently developed substrate technologies engineered to circumvent some of the traditional limitations that existed just a few years ago and that severely impacted the RF performance of CMOS devices in SOI. These SOI starting substrates are now commonly available in very high volumes from a number of substrate suppliers, which has been the key enabler for RFSOI technologies to achieve wide acceptance by the design community.

RFSOI technologies on silicon substrates (both standard and high resistivity) have been explored since the late 1980s. As SOI processes and substrates matured, several research groups demonstrated the substantial improvements in cross talk, isolation, and power dissipation over conventional bulk digital and analog technologies were possible. Honeywell,[2] Motorola,[3] and Harris[4] as well as many in academia in the United States and several other research laboratories in Japan published several leading articles on RFSOI on silicon substrates, identifying the benefits and trade-offs for RF building blocks inherent in a thin film silicon layer over oxide. Motorola published extensively in the mid-1990s on the demonstration of several key RF building blocks such as switches, LNAs, mixers, and power amplifiers built on RFSOI technologies,[5] also demonstrating voltage charge pumps and digital control with a high degree of integration. Similar development efforts were also being conducted in Asia and Europe at the same time. Honeywell, however, is generally credited with commercially offering the very first RF switches and RF components built on RFSOI on silicon substrates in the late 1990s, with many of the design features that are used to this day in our 3G/4G switches; the company also pioneered the offering of the RFSOI-on-silicon process as a general foundry option and very early realized the potential benefits of high-resistance substrates and cross talk isolation for superior RF performance.* On the cellular phone side, Peregrine Semiconductor is generally credited with achieving the first SOI design win in a mobile handset in the early 2000s, with its own proprietary silicon on sapphire (SOS) technology.[6]

In order for RFSOI-on-HR silicon technologies to compete—and eventually replace—other conventional switching technologies such as III–V PHEMT (pseudomorphic high electron mobility transistor) and SOS solutions, three completely separate technological challenges had to be resolved in order to bring about the necessary improvements in RF performance, yield, and cost that would eventually make RFSOI-on-HR the leading technology that it is today:

1. *SOI device yield*: RFSOI-on-HR silicon technologies typically employ silicon device regions with layer thicknesses in the 800–2000 Å range; such thin layers must be controlled very uniformly across the wafer and must be free of defects and other issues that negatively affect CMOS device characteristics and reliability. Conventional SOI wafer techniques involving wafer bonding and thinning of the silicon epitaxy layer proved to be extremely challenging for layer thicknesses under 1000 Å. An alternative technique called SIMOX yielded the first generation of RFSOI substrates used in early demonstrations in the 1990s, where the BOX (buried oxide) layer was built by the thermal oxidation of a high-density oxygen implant. Whereas the SIMOX technique appeared on paper to offer a significant improvement in cost and simplicity over other SOI bonding techniques, it was quickly determined to be fraught with high levels of defectivity and severe problems with wafer uniformity and device yield. The "SMART-CUT" process developed by SOITEC in France is credited with providing the key needed

---

* CMOS SOI Technology, Honeywell Technical Note, http://www.techonline.com/electrical-engineers/ education-training/tech-papers/4134069/SOI-CMOS-Technology-for-RF-System-on-Chip- Applications, October 2001; see also http://pdf.datasheetarchive.com/indexerfiles/Datasheets-IS15/ DSA00283852.pdf for datasheets for commercial RFSOI switch products from Honeywell introduced in 2001.

technological leap in SOI substrate technology; SMART-CUT SOI sub-strates very quickly were shown to solve the nagging problems associated with other SOI substrate approaches.[7]

2. *Lack of stable/low-cost high-resistivity silicon substrate*: Similar to RF power switches built previously in III–V technologies, a basic building block for switches and tuners is the ability to cascade (stack) several FET devices in series to realize large OFF state RF voltages. This is only possible to achieve with a high-resistivity substrate that provides the necessary isolation between different devices that comprise the switch branch. This concept will be explained in a later section of this chapter. High substrate resistivities are readily available in GaAs and sapphire substrates with values in the many $M\Omega \cdot cm$ range, but it took significant effort by the SOI substrate providers to develop the low-cost >1 $k\Omega \cdot cm$ silicon substrates using conventional CZ sil-icon wafer equipment used ubiquitously by the silicon SOI industry today.* Although silicon HR substrates today are only available with resistivities many orders of magnitude lower than III–V and sapphire substrates, the 1 $k\Omega \cdot cm$ range has been shown to be adequate for the frequency ranges and power levels used in the mobile handset industry. On the other hand, silicon substrates have a much higher thermal conductivity value than sapphire and GaAs, a property that is particularly important for applications that employ flip-chip packaging or that integrate power amplifiers. A deep understanding of the effects of oxygen interstitial vacancies and necessary thermal treat-ments on resistivity required several years of applied research by the many silicon substrate vendors. Today, these high-resistivity (HR) 200 mm SOI substrates offer substantial cost advantage over competing substrate technol-ogies such as GaAs and sapphire and have already been successfully dem-onstrated in 300 mm wafer diameters for future RFSOI platforms. Starting wafer resistivities in the 5,000–10,000 $k\Omega \cdot cm$ will soon be available in very high volumes with a cost-effective large volume supply chain.

3. *Substrate-induced harmonics*: Early demonstrations of RF switches in RFSOI-on-HR silicon demonstrated much higher harmonic and intermodu-lation distortion than expected from a simple analysis of the ON and OFF transistor characteristics. Independently, a few research groups concluded in the early 2000s, that the origin of such deleterious harmonics was due to the silicon substrate itself.[9] The RF linearity issue due to the HR substrates took significantly longer to be correctly understood and resolved. In paral-lel, independent RFSOI research groups reported seeing rather strange and variable RF linearity and harmonic data at higher powers, despite having impeccable DC/AC parametric data for the CMOS active devices. Further analysis led to the conclusion that the interface between the BOX region and the high-resistivity silicon handle substrate was the source of such del-eterious effects. Two very distinct solutions for the substrate-induced lin-earity degradation were developed and are now deployed in the industry in very large production scale:

---

* See, for example, Rong et al.[8]

- *Trap-rich* polysilicon regions (see Figure 2.6a) were introduced at the BOX/silicon handle interface and shown to play a dramatic role in controlling the nonlinear harmonics created by the HR substrate.[10] Originated by the pioneering work of Professor Raskin at the Catholic University of Louvain (Belgium) and perfected by the interaction between industry and SOI substrate vendors such as SOITEC, the trap-rich technique is now commonplace in the wireless industry and allowed several new silicon foundries to develop simple adaptations of their existing CMOS and SOI processes to create an RFSOI technology offering.
- *Harmonic suppression process (HSP) techniques*: Developed by the SOI technology team at IBM[11] (see Figure 2.6b), the HSP process consisted of a series of damaging implants, trenches, and proprietary treatments to minimize the contribution of the BOX/handle substrate to RF linearity. IBM has recently announced the shipment of over five billion RFSOI switch devices, demonstrating the efficacy and high-volume stability of its harmonic suppression process.

As a major advantage when compared with III–V and sapphire-based technologies, RFSOI-on-HR silicon technologies typically employ the exact same back-end-of-line metallization, bumping, thinning and copper pillar, and other post-processing technologies utilized in conventional silicon bulk technologies with no needed modifications or added costs. Several leading wireless companies today ship millions of RFSOI-on-HR silicon components per month with extremely high production

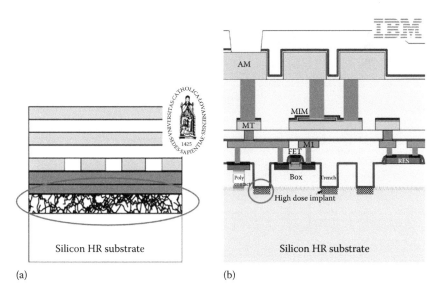

(a)   (b)

FIGURE 2.6 Two approaches to solve substrate-induced harmonics commonly used in RFSOI on HR silicon technologies: (a) A buried polysilicon "trap-rich" layer is employed and (b) a combination of damaging implants and trenches are employed for the same objective.

volumes.[12,13] The industry is currently gearing up for migration to 300 mm production in the near future, clearly showing its maturity and importance in the wireless product landscape.

2.3   SWITCH BRANCH DESIGN

The N-channel field effect transistor (NFET) switch branch is the basic building block for switches and tunable/reconfigurable circuits built in RFSOI on-HR-silicon technologies. The typical switch branch consists of a combination of stacked NFET devices with MIM capacitors and high-value resistors connected to the gate and body terminals, built on a thin (800–2000 Å) silicon layer over a buried oxide (BOX) region over a high-resistivity substrate (see Figure 2.8). Currently, all RFSOI-on-HR silicon technologies used for RF front-end applications used partially depleted (PD) SOI configurations, as the PD option seems to provide the best trade-off for power handling and reduction of harmonics through appropriate biasing of the body-contact terminal. Fully depleted (FD) SOI technologies are, of course, also available as foundry options typically in the 300 mm wafer diameter, utilizing much thinner device, epitaxy thicknesses, and much more aggressive gate lithography. FD SOI technologies remain mostly focused on digital low-power VLSI applications and at this time have not found utilization in RF large power switch designs.

Figure 2.7 is an illustration of these different device regions in a commercially available RFSOI technology [14].

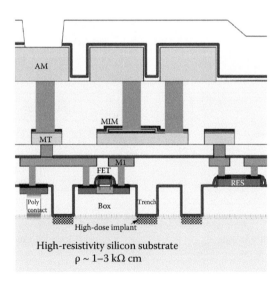

FIGURE 2.7   IBM 7RFSOI NFET transistor cross section, also showing the implementation of an MIM capacitor and polysilicon resistor as well as the harmonic suppression process (HSP) trenches and implants.

The substrate resistivity must be high enough with respect to the frequency of operation to allow the substrate to behave strictly as a capacitive network, which is easily accomplished with today's available values in the range of 1–3 $k\Omega \cdot cm$. The stacked group of the RFSOI NFETs is the basic building block for essentially any application in this field. At its core, lies the basic thick-gate NFET switch device. Today, nearly all RFSOI on-HR silicon switch and tuner solutions use a 0.25 um NFET output (also known as "thick gate") device built on either a 0.18 um or a 0.13 um SOI CMOS foundry technology, employing many different layout configurations to optimize the RF performance of the switch branch. The thinner gate oxide option in the technology is typically omitted to minimize costs and cycle time, although with the current drive to integrate low-noise amplifiers (LNAs) in front-end modules, more foundries are offering the thinner gate FET as a device option for such applications. PFET devices are typically not utilized in RF switch branches due to their inherently poorer carrier mobility and higher RDSON characteristics, although PFETs generally play a very important role in the implementation of integrated power controllers for RF applications.

RFSOI switch branches are typically biased using gate switching between +VDD/VNEG in the ON/OFF states respectively, where VDD and VNEG are high and low voltage levels supplied by the power management block of the IC. Typical VDD and VNEG voltages used in today's RFSOI applications employ values around ±2.5 V, respectively. The drain and source voltages are biased at 0 V. Conventionally, body-contacted FETs are used, where the body would be biased through high-value resistors and typically be switched between 0 V/VNEG in the ON/OFF states respectively.

2.3.1 $R_{ON}C_{OFF}$

A key figure of metric (FOM) utilized in RFSOI technologies is the product of the ON resistance of a switch device with its respective OFF state capacitance ($R_{ON}C_{OFF}$), generally expressed in femtoseconds. When associated with a specified device breakdown value and power handling, the $R_{ON}C_{OFF}$ product is a valid predictor of the device performance in an RF switch application. RFSOI-on-HR silicon technologies currently report $R_{ON}C_{OFF}$ figures of merit in the 240–150 fs range. Several new improvements are currently being developed to continuously decrease the switch FOM, such as the introduction of all copper metallization, finer lithography for improved gate pitches, and several other proprietary techniques will reduce this figure to the 110–150 fs range in 2015–2016 time frame.* It is important to keep in mind that this FOM is only meaningful when comparing RFSOI technologies of similar RF power handling and DC breakdown. Constant improvements in processing techniques and materials will undoubtedly push RFSOI figures of merit to the sub-100 fs range in 2017–2018 time frame.

In the OFF state, the power handling of a given switch branch is determined by the number of stacked NFETs in series, as summarized in Figure 2.8.

---

* See, for example, http://www.marketwatch.com/story/peregrine-semiconductor-ships-first-ultracmos-10-production-units-2014-05-06.

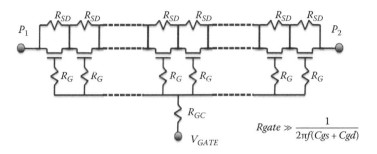

FIGURE 2.8 Typical RFSOI switch branch topology, showing N× stacked switch FETs, each DC-biased via a high-value gate resistor according to the frequency of operation and the technology OFF state capacitance. High-value resistors may also be employed across the source–drain terminals of the device for biasing and voltage distribution.

The goal of the combination of stacked switch devices and gate and body resistors is to equally divide a large RF voltage into smaller RMS values consistent with the breakdown characteristics of each individual transistor. This stacking property of RFSOI NFET devices is, of course, identical to the strategies utilized in the past by III–V switch designers as summarized in a classic paper by M. Shifrin.[14] Shifrin also highlighted the need for high-value DC biasing gate resistors, with the required resistor value being a function of the RF of operation and the OFF state capacitance associated with the given switch branch transistor. In RFSOI switches and tuner applications, typical stack numbers may vary between $N = 6$ and $N = 24$, depending on the magnitude of the RMS voltage the branch must withstand in the OFF state. The switch designer must pay close attention to the harmonics of the OFF state device under the highest rated RF power levels and mismatched, as harmonics levels will start increasing dramatically prior to the onset of voltage breakdown.

Once properly designed for a given power level/insertion loss/isolation requirement, switch branches become the fundamental building block of a myriad of different RF applications such as power switches, programmable attenuators, impedance tuners, tunable networks, and programmable filters, employing a combination of series and shunt branches and other internal and external passive and active components.[15,16] Figure 2.9 illustrates the block diagram of a typical single pole–four throw (SP4T) antenna switch application with the different series and shunt branches and its associated small-signal equivalent circuit. The shunt branches are key elements that provide necessary port-to-port isolation and the series branches are the blocks that connect to the antenna (the single pole) and hence are the key elements that determine the insertion loss of each branch.

Typical RFSOI on-HR silicon switch products in the market today integrate all of the necessary power controllers and digital logic blocks needed for a given application. Most RFSOI technologies offered by foundries also offer "e-fuse" elements, which are one-time programmable elements useful for calibration or identification of a switch or tuner IC.

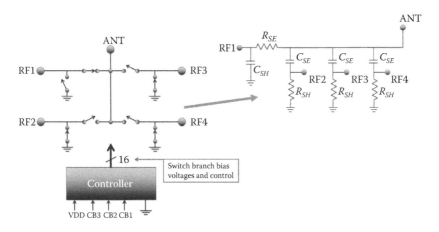

FIGURE 2.9 A typical block diagram of a single pole–four throw (SP4T) RFSOI switch showing the combination of switch branches in a given application and its small-signal equivalent circuit. The power handling of the switch is a function of the configuration and design of each of the individual OFF state switch branches. Nearly all RFSOI circuits employ integrated biasing controllers.

2.3.2 PSP Electrical Compact Model

RF switch designers today rely a great deal on compact models to predict the electrical behavior of a given RFSOI switch branch with respect to small-signal characteristics such as insertion loss and isolation, as well as important large signal behavior such as second- and third-order harmonics and intermodulation distortion products.[13] The PSP model is used ubiquitously by RFSOI designers to model the actual NFET device in partially depleted (PD) SOI technologies, used in conjunction with a capacitive/resistive network that models the complex nature of the substrate that ties the different stack devices together. The PSP model was jointly developed by the Pennsylvania State University and Philips Research and is a surface-potential-based MOS model containing many relevant effects for RF switch simulations such as mobility reduction, velocity saturation, drain-induced barrier lowering (DIBL), gate and body currents, lateral doping gradient effects, and so on. The PSP model is generally favored for RF design over the more standard BSIM-SOI models, due to the inherent poor harmonics and intermodulation agreement in the BSIM-vintage models with measured results; BSIM models also tend to exhibit problematic discontinuities in the derivatives of transconductance where switches typically operate, resulting in poor convergence and agreement. PSP-SOI surface-based models continue to be optimized by the modeling community and are expected to continue to be the preferred model by the RFSOI foundries and designers alike.[17]

For optimal agreement of RFSOI switch branches, it is also highly recommended that the designer employ parasitic-extraction (PEX) programs to model all additional resistive and capacitive effects of the metallization that connect the different switch branches and that are not comprised in the original PSP model. Nearly all commercial

RFSOI on-high-HR silicon technologies today provide the designer with a complex process design kit (PDK) that includes design rule checking (DRC), layout versus schematic (LVS) checking, and PEX, in addition to extensive digital/analog/ESD and effuse library elements.

2.4 PACs: PROGRAMMABLE ARRAY OF CAPACITORS

RFSOI programmable array of capacitors (PACs) are another common RF building block, being employed in very high volumes in the cellular handset for impedance and aperture tuning at the antenna level and for a myriad of other applications where tunability and reconfigurability are desired. PACs are essentially a switchable bank of MIM capacitors, which can be turned ON and OFF by an associated series RFSOI switch branch described in the previous section.

MIM capacitors available in RFSOI-on-HR silicon technologies have very respectable Quality factors (Q) by themselves ($Q > 100$ at 2 GHz); typically there are two different MIM capacitor options in a given RFSOI technology with a thin silicon nitride layer for high-density and a thicker silicon nitride layer for higher-voltage applications. The effective equivalent series resistance (ESR) and associated Q of the switched capacitor will be dominated by the ON resistance of the series switch branch, so selecting the appropriate stack number and W (width) of the series branch is extremely important as there will be a trade-off between Q, power handling and the physical silicon real estate area allocated to the switched capacitor. For antenna tuning applications, typically the RFSOI designer will employ very high stack numbers due to the very high RMS voltage requirements present in an antenna environment; RON values in the 0.3–0.5 Ω can still be concurrently obtained by the appropriate choice of device width, yielding typical Q factors of 40–60 at $f = 2$ GHz for a switched 4 pF MIM capacitor. RON values of tenths of ohms—necessary, for example, for a $Q = 200$ for the same 4 pF capacitor—are unfeasible in a switched SOI capacitor configuration due to the required area of the stacked switch and associated metal losses.

The most utilized type of programmable array of capacitors (PAC) is a parallel array of binary weighted capacitors in series with high dynamic range RF switches (Figure 2.10a). A series array of capacitors (Figure 2.10b) can also be constructed, with quite different Q and dynamic range characteristics versus tuning conditions when compared to the parallel array. The two configurations offer the RFSOI designer different trade-offs with respect to Q and voltage range. The parallel array's signal handling capability is essentially independent of tuning and is ideally suited for use in parallel resonant circuits. The series array's signal handling, on the other hand, is inversely proportional to capacitance and therefore lends itself to the tuning of series resonances. A combination of both parallel and series approaches gives the RFSOI designer a versatile toolbox of tunable and reconfigurable elements for RF design.

The Quality factor Q of a PAC is a function of tuning range C_{max}/C_{min} (maximum and minimum switched capacitor values) and the $R_{ON}C_{OFF}$ FOM of the RF switch process being used, as indicated in Equation 1.[18] By restricting the PAC's tuning range, Qs greater than 100 are obtainable using the most advanced RFSOI processes available in the industry today.

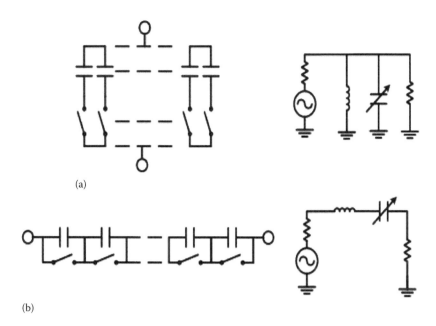

(a)

(b)

FIGURE 2.10 Programmable array of capacitors (PACs) arranged in (a) parallel or (b) series configurations. Each configuration has different trade-offs in Quality factor Q and power handling capabilities.

$$Q_{min} = \frac{1}{((C_{max}/C_{min})-1)\cdot R_{ON}\cdot C_{OFF}} \qquad (2.1)$$

The Q_{min} equation assumes that the C_{min} is solely determined by the C_{OFF} of the switches and that the FETs have been sized to meet the target C_{min}. In moderate Q applications the target C_{min} can be achieved by placing a high-quality MIM or MOM capacitor in parallel to a PAC structure. This gives only a modest Q improvement compared to scaling the FET switch size but has the benefits that the C_{min} correlates closer to C_{max} and the die area is not dominated by a huge switch.

The design of a PAC consists in trading off chip area for the necessary capacitance tuning ratio and effective Quality factor for a maximum RF voltage waveform; obviously the better the $R_{ON}C_{OFF}$ figure of merit for a given RFSOI technology, the lesser the impact of the switch branch. This trade-off between capacitor tuning ratio (C_{max}/C_{min}) and switch branch resistance and technology figure of merit $R_{ON}C_{OFF}$ was summarized in a classic RFSOI paper by Whatley et al.,[18] as depicted in Figure 2.11. Clearly, moving forward with further improving the $R_{ON}C_{OFF}$ figure of merit will play a very positive role in minimizing the trade-off of Q versus tuning ratio for future RFSOI on high-resistivity silicon technologies.

MIM capacitors themselves are capable of generating second- and third-order harmonics and also having significant temperature-dependent effects, although these are typically below the levels of harmonics created by the series RFSOI switch blocks.

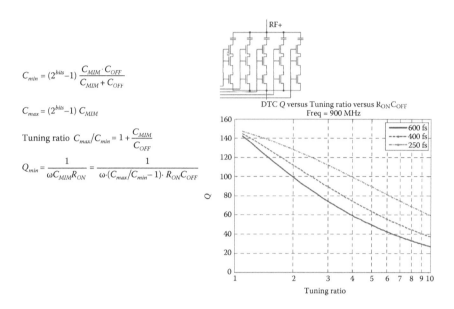

$$C_{min} = (2^{bits}-1)\,\frac{C_{MIM}\cdot C_{OFF}}{C_{MIM}+C_{OFF}}$$

$$C_{max} = (2^{bits}-1)\,C_{MIM}$$

$$\text{Tuning ratio}\ \ C_{max}/C_{min} = 1 + \frac{C_{MIM}}{C_{OFF}}$$

$$Q_{min} = \frac{1}{\omega C_{MIM}R_{ON}} = \frac{1}{\omega\cdot(C_{max}/C_{min}-1)\cdot R_{ON}C_{OFF}}$$

FIGURE 2.11 Trade-off of Quality factor Q versus tuning ratio of an SOI switched MIM capacitor for different $R_{ON}C_{OFF}$ figures of merit. (After Whatley, R. et al., RF front-end tunability for LTE handset applications, *Compound Semiconductor Integrated Circuit Symposium (CSICS)*, 2010[18].)

Nonlinear characteristics of MIM capacitors available in silicon RFSOI technologies are generally attributed to traps generated in the silicon nitride films' chemical vapor deposition (CVD) process and are generally included in the large-signal electrical model in the technology process design kit (PDK).[19] Because these MIM nitride films are introduced in the back-end portion of the RFSOI process, there are strict limitations on the deposition temperatures of dielectrics that impact the overall dielectric film quality. The second-order effects due to MIM capacitor nonlinearity can be cancelled by design approaches employing a combination of antiparallel capacitors. The third-order harmonic generation in a MIM capacitor is more difficult to compensate for in a given design; as done for RFSOI switch branches, a designer will typically stack a higher number of MIM series capacitors to reduce the effective RMS voltages across each MIM capacitor element, although with serious impact in the die area utilized for the PAC.

2.4.1 PAC DYNAMIC RANGE

The dynamic range and signal handling of a PAC (programmable array of capacitors) is determined by a combination of different design trade-offs and inherent characteristics of the RFSOI process itself. Most importantly, dynamic range and power handling of a PAC are determined by the degree of stacking used in the switch design, the FOM of the RFSOI process used, the nonlinearity of the MIM capacitors, and eventually substrate distortion caused by the HR silicon/BOX interface. The RFSOI designer should explore different parallel (Figure 2.10a) and series (Figure 2.10b) configurations or combinations of the two for optimal Q at a given power handling. Theoretically, for a properly designed switch, doubling the degree of stacking (the number of NFET switch branches in series) in conjunction with doubling the width of the FET branches will result in a 6 dB increase in power handling without any accompanying degradation of the PAC Q. Antenna tuners based on high stacking numbers are now commonplace in modern handsets, with stacking numbers in the $N = 24$–36 range being reported, capable of meeting RMS voltages greater than 60 V. This ideal scaling of power handling with stack number, however, starts deviating from the ideal 6 dB slope after stack numbers of $N = 20$ or more due to uneven voltage division in the switch branches at very high RMS voltages and needs to be taken into account experimentally by the designer as this effect is currently not something that can be simulated by the PDK models.

2.5 RFSOI APPLICATIONS IN THE CELLULAR HANDSET

There has been a stunning growth in the number of RFCMOS on high-resistivity silicon designs that have reached very large volume production in wireless applications during the 2013–2015 time frame. These designs cover a multitude of different switch configurations and tunable/reconfigurable products targeting 3G and 4G cellular applications and are available both as discrete individual components and as a part of more complex RF modules encompassing a combination of power amplifiers, CMOS controllers, filters, and IPD passive devices. RFSOI-on-HR silicon technology has entered the mainstream of RF technologies and largely replaced conventional III–V PHEMT switch offerings in the cellular space due to its high degree of integration, small size, lower cost, and excellent RF performance. RFSOI-on-HR silicon switches are now available commercially from a number of different vendors using SOI technologies offered at the foundry level; currently production levels are estimated to exceed 40,000 wafers per month in 2014.

2.5.1 STANDALONE PACS

Individual RFSOI PACs are useful and versatile components that find use in many tuning applications. These components are offered in a number of different binary configurations (from 4-bit to 32-bit binary tuning) depending on the necessary minimum capacitor resolution needed for the application. A 4-bit binary weighted PAC design that provides a 0.5–5 pF tuning range is shown in Figure 2.12.[1] It contains all of the digital/analog circuitry, including the serial interface (GPIO or RFFE), charge pump, ESD network, and switch drivers to provide extremely small and cost-effective tuning solutions. This very compact and highly integrated design and other similar configurations are being shipped in extremely high volume in antenna applications in the cellular handset market.

The measured performance for a 4-bit PAC on using an SOI process with a 220 fs FOM (IBM 7RFSOI) is shown in Figure 2.13. The device's Q falls short of the Q_{min} predicted by Equation 1. This deviation from the ideal is caused by a combination of the loading of the biasing network, routing resistance, and parasitic capacitances that reduce the net C_{max}/C_{min} ratio. These effects can, however, be readily simulated by

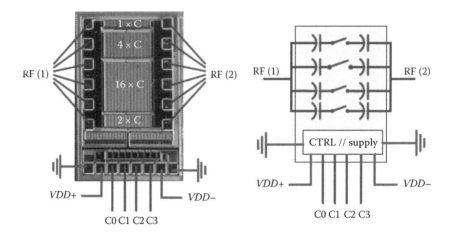

FIGURE 2.12   Tunable array of capacitor built with RFSOI switch branches and MIM capacitors. (From Costa, J., *IEEE Microwave Mag.*, 15(7), S61, November–December, 2014.)

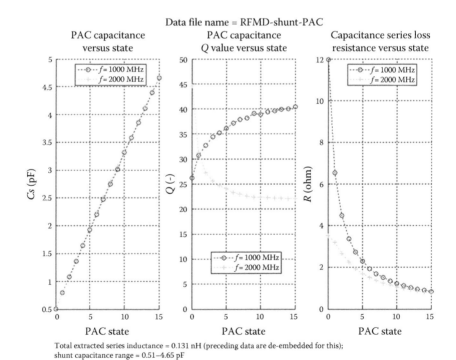

FIGURE 2.13   A 4-bit PAC capacitance, Q and series resistance versus tuning state.

	Shim [27]	Natarajin [28]	Baxter [29]	Im [30]	Granger-Jones [23]
Tuning ratio	5	10	5.1	4.7	9.4
C_{min} (pF)	0.2	0.13	0.9	1	0.5
C_{max} (pF)	1	1.3	4.6	4.7	4.7
Step size (pF)	CT	0.125	0.12	0.25	0.28
Q-factor at C_{min}					
1 GHz	>100		42	47	26
2 GHz	>100		50	50	41
Q-factor at C_{max}					
1 GHz	100	160	29	34	45
2 GHz	50	87	13	15	23
Technology	MEMS	MEMS	SOS	SOI	SOI

FIGURE 2.14 A comparison of RF PACs implemented in different technologies.

adding parasitic-extraction (PEX) generated networks, which will correct for these parasitic networks.

A recent comparison of published PAC performance[21] is shown in Figure 2.14, including results reported for PACs, implemented in alternative technologies such as microelectromechanical systems (MEMS) and silicon on sapphire (SOS).[20–23]

2.5.2 RFSOI IMPEDANCE TUNERS

An impedance tuner is, by definition, a network that attempts to provide an optimal impedance match between antenna and the RF front-end blocks; an aperture tuner, on the other hand, serves a related but different function, in that it provides the optimal impedance between RF front end and antenna to maximize transmitted RF power. Both of these important antenna tuning functions are easily synthesized in modern RFSOI technologies and are now being shipped in very high volumes in the mobile handset. RFSOI technologies offer the added benefit that a low-impedance bypass can always be added to the tuner by the simple introduction of a series switch branch, which offers additional versatility and performance improvement.

RFSOI impedance tuners are realized by a combination of RF switch branches, programmable array of capacitors (PACs), and internal/external components to realize high-Q impedance tuning networks that optimize the power delivered to a cellular handset antenna under a number of different environmental conditions. These tunable/reconfigurable networks can employ either series or parallel PAC tuning elements depending on the nature of the resonating structure of the application. Typical impedance tuner solutions include a combination of external high-Q inductors working in conjunction with the internal tunable MIM capacitors and a combination of RF switches to efficiently synthesize a number of different impedance states that optimize the impedance match of a cellular antenna to the other components of the RF front end.

Die photo

Application board photo

FIGURE 2.15   Antenna tuner implemented with a combination of RFSOI die and external high-Q inductors. The die area is 2.4 × 2.6 mm and the PAC covers a capacitor range between C_{min} = 0.5 pF and C_{max} = 4.6 pF.

Figure 2.15 depicts a broadband SOI impedance tuner that has been utilized in very high volume in several leading 4G handsets, which covers a capacitor range of C_{min} = 0.5 pF to C_{max} = 4.6 pF, for a die size of 2.4 mm × 2.6 mm.

This particular PAC design serves as the configurable element in an antenna impedance tuner designed to cover the frequency range between 0.7 and 2.2 GHz. In this particular application, the overall Q of the entire impedance tuner ended up being limited just as well by the Q of the external inductors used along with the PAC (typical inductor Q's at 1 GHz hover in the 40–55 range).

The die is implemented in a popular RFSOI-on-HR silicon technology (IBM 7RFSOI). The impedance tuner depicted performs very well in matching and optimizing the varying impedance characteristics to the RF front-end module for a given antenna. The net gain or loss of this particular RFSOI antenna tuner at 895 MHz is illustrated in Figure 2.16; similar charts are available for this same tuner in different frequency band.

The RF antenna tuner gain chart of Figure 2.16 is a very useful tool to visualize the performance of a given RF tuner in a given frequency band under a different set of load characteristics. This particular chart shows that the insertion loss of the tuner is about 0.56 dB; it is also very useful to determine the "breakeven" state, which corresponds to the G_{TUNER} = 0 dB where the impedance tuning essentially compensates for the insertion loss of the tuner itself. Outside of the "breakeven" circle, there is essentially a net gain in performance due to the RFSOI tuner, so obviously, the designer seeks the optimal combination of RFSOI tuner and external components to realize the smallest possible "breakeven" circle.

The antenna tuner depicted in Figure 2.16 shows an insertion loss of 0.56 dB at 1:1 VSWR. This loss, however, is recovered very quickly with VSWRs in the ≥2 range, and as VSWR increases, the tuner realizes net positive gain for the overall network showing a clear improvement in typical antenna implementations.

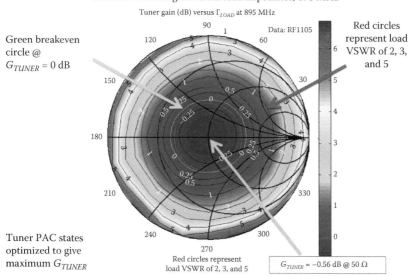

Tornado 2: Tuner gain versus load impedance; 895 MHz

Tuner gain (dB) versus Γ_{LOAD} at 895 MHz

Green breakeven circle @ G_{TUNER} = 0 dB

Red circles represent load VSWR of 2, 3, and 5

Data: RF1105

Tuner PAC states optimized to give maximum G_{TUNER}

Red circles represent load VSWR of 2, 3, and 5

G_{TUNER} = −0.56 dB @ 50 Ω

FIGURE 2.16 Low-band (895 MHz) impedance tuner performance. The colors in the circle represent the effective gain/loss of the combined impedance tuner and antenna. At VSWR's levels approaching 2, the tuner starts providing a net performance gain for the overall circuit, and at higher VSWR levels, the gain improvement becomes very significant.

2.6 FUTURE TRENDS IN RFSOI-ON-HR SILICON TECHNOLOGIES

4G cellular switching systems will continue to drive enhancements in both RF switch linearity and harmonic/intermodulation distortion on next-generation RFSOI-on-HR silicon technologies. Some of these additional performance improvements will come from proprietary circuit design and optimization being done by the major players in this market. In addition to these techniques, there will definitely be significant push in the technology front to optimize the fundamental characteristics of the RFSOI technology.

Several different trends are being investigated in the industry, including handle substrate resistivity optimization and linearization, migration to 300 mm wafer diameters, as well as a push to achieve the best possible device size reduction—and its accompanying reduction in harmonics and intermodulation products. 300 mm wafer processes make it available finer lithography modules, copper metallization and low-K dielectrics, all promising candidates to further reduce the FOM of future RFSOI technologies. Figures of merit of R_{ON}–C_{OFF} will likely approach values in the 110–150 fs range in the 2015–2016 time frame with improvements in fabrication and device design, with expectations of FOM < 100 fS in production in the 2017–2018 time frame. The constant drive for performance must also be balanced with costs and total solution die size, but overall, these are the two areas where RFSOI-on-HR silicon substrates has come to play a dominant role.

ACKNOWLEDGMENTS

The author would like to acknowledge the invaluable technical contributions of the following engineers in the preparation of the materials covered in this chapter: Mike Carroll, Phil Mason, Dan Kerr, Christian Iversen, Ali Tombak, Beth Glass, Nadim Khlat, Marcus Granger-Jones, and Eddie Spears.

REFERENCES

1. Costa, J., Passing the plateau of productivity: Development of RFSOI technologies on HR silicon substrates for reconfigurable wireless solutions, *IEEE Microwave Magazine*, 15(7), S61–S73, November–December 2014.
2. Liu, S.T., Fechner, P., Balster, S., Dougal, G., Sinha, S., Chen, H., Shaw, G., Yue, J., Jenkins, W.C., and Hughes, H.L., A 5 nm nitrided gate oxide for 0.25/spl mu/m SOI CMOS technologies, *IEEE Transactions on Nuclear Science*, 46(6), 1824–1829, December 1999.
3. Huang, W.M., Ngo, D., Babcock, J.I., Shin, H.C., Welch, P., Racanelli, M., Foerstner, J., Ford, J., and Cheng, S., TFSOI complementary BiCMOS technology for low power RF mixed-mode applications, *IEEE Custom IC Conference*, 1996.
4. Krull, W.A., Buller, J.F., Rouse, G.V., and Cherne, R.D., Electrical and radiation characterization of three SOI material technologies, *IEEE Circuits and Devices Magazine*, 3(4), 20–26, July 1987, doi: 10.1109/MCD.1987.6323129.
5. Lovelace, D., David, N., Costa, J., and Camilleri, N., Silicon MOSFET technology for RF ICs, *Sixth IEEE International Symposium on Personal, Indoor and Mobile Radio Communications, 1995 (PIMRC'95). Wireless: Merging onto the Information Superhighway*, Toronto, Ontario, Canada, vol. 3, p. 1238, September 27–29, 1995.
6. Kelly, D., Brindle, C., Kemerling, C., and Stuber, M., The state-of-the-art of silicon-on-sapphire CMOS RF switches, *IEEE Compound Semiconductor Integrated Circuit Symposium (CSIC '05)*, 2005.
7. Wittkower, A., Auberton-Herve, A., and Maleville, C., SMART-CUT(R) technology for SOI: A new high volume application for ion implantation, *Conference on Ion Implantation Technology*, Alpbach, Austria, pp. 269–272, 2000.
8. Rong, B., Burghartz, J.N., Nanver, L.K., Rejaei, B., and Van der Zwan, M., Surface-passivated high-resistivity silicon substrates for RFICs, *IEEE Electron Device Letters*, 25(4), 176–178, April 2004.
9. Kerr, D.C., Gering, J.M., McKay, T.G., Carroll, M.S., Roda Neve, C., and Raskin, J.-P., Identification of RF harmonic distortion on Si substrates and its reduction using a trap-rich layer, *IEEE Topical Meeting on Silicon Monolithic Integrated Circuits in RF Systems, 2008 (SiRF 2008)*, pp. 151–154, January 23–25, 2008.
10. Ben Ali, K., Roda Neve, C., Gharsallah, A., and Raskin, J.-P., RF SOI CMOS technology on commercial trap-rich high resistivity SOI wafer, *IEEE International SOI Conference (SOI)*, pp. 1–2, October 1–4, 2012.
11. Botula, A., Joseph, A., Slinkman, J., Wolf, R., He, Z.-X., Ioannou, D., Wagner, L. et al., A thin-film SOI 180nm CMOS RF switch technology, *IEEE Topical Meeting on Silicon Monolithic Integrated Circuits in RF Systems, 2009 (SiRF '09)*, pp. 1–4, January 19–21, 2009.
12. Carroll, M., Kerr, D., Iversen, C., Tombak, A., Pierres, J.-B., Mason, P., and Costa, J., High-resistivity SOI CMOS cellular antenna switches, *Annual IEEE Compound Semiconductor Integrated Circuit Symposium, 2009 (CISC 2009)*, pp. 1–4, October 11–14, 2009.

13. Lee, T.-Y. and Lee, S. Modeling of SOI FET for RF switch applications, *Radio Frequency Integrated Circuits Symposium (RFIC), 2010 IEEE*, pp. 479–482, May 23–25, 2010.

14. Shifrin, M.B., Katzin, P.J., and Ayasli, Y., *IEEE Transactions on Microwave Theory and Techniques*, 37, 2134–2141, 1989.

15. Blaschke, V., Zwingman, R., Hurwitz, P., Chaudhry, S., and Racanelli, M., A linear-throw SP6T antenna switch in 180 nm CMOS thick-film SOI, *2011 IEEE International Conference on Microwaves, Communications, Antennas and Electronics Systems (COMCAS)*.

16. Wang, D., Wolf, R., Joseph, A., Botula, A., Rabbeni, P., Boenke, M., Harame, D., and Dunn, J. High performance SOI RF switches for wireless applications, *International Conference on Solid-State and Integrated Circuit Technology (ICSICT)*, 2010.

17. Wu, W., Li, X., Gildenblat, G., Workman, G., Veeraraghavan, S., McAndrew, C., van Langevelde, R. et al., PSP-SOI: A surface potential based compact model of partially depleted SOI MOSFETs, *IEEE Custom Integrated Circuits Conference, 2007 (CICC '07)*, pp. 41–48, September 16–19, 2007.

18. Whatley, R., Ranta, T., and Kelly, D., RF front-end tunability for LTE handset applications, *Compound Semiconductor Integrated Circuit Symposium (CSICS)*, 2010.

19. Babcock, J.A., Balster, S.G., Pinto, A., Dirnecker, C., Steinmann, P., Jumpertz, R., and El-Kareh, B., Analog characteristics of metal-insulator-metal capacitors using PECVD nitride dielectrics, *IEEE Electron Device Letters*, 22(5), 230–232, May 2001.

20. Shim, Y., Wu, Z., and Rais-Zadeh, M., A high-performance, temperature-stable, continuously tuned MEMS capacitor, *2011 IEEE 24th International Conference on Micro Electro Mechanical Systems (MEMS)*, pp. 752–755, January 23–27, 2011.

21. Natarajan, S.P., Cunningham, S.J., Morris, A.S., and Dereus, D.R., CMOS integrated digital RF MEMS capacitors, *2011 IEEE 11th Topical Meeting on Silicon Monolithic Integrated Circuits in RF Systems (SiRF)*, pp. 173–176, January 17–19, 2011.

22. Baxter, B., Ranta, T., Facchini, M., Dongjin, J., and Kelly, D., The state-of-the-art in silicon-on-sapphire components for antenna tuning, *2013 IEEE MTT-S International Microwave Symposium Digest (IMS)*, pp. 1–4, June 2–7, 2013.

23. Im, D. and Lee, K., Highly linear silicon-on-insulator CMOS digitally programmable capacitor array for tunable antenna matching circuits, *IEEE Microwave and Wireless Components Letters*, 23(12), 665–667, December 2013.

24. Larry Schumacher, Challenges for radios due to carrier aggregation requirements, *International Wireless Industry Consortium (IWPC) Workshop*, Intel Mobile Corporation, November 6, 2012.

3 BST Technology for Mobile Applications

James G. Oakes and David W. Laks

CONTENTS

3.1 INTRODUCTION

Mobile devices are used in a growing number of consumer applications driving up the demand for both voice and data from carriers. As a result of this demand, the carriers must provide high data rates for their customers, and the mobile phone OEMs have to provide cost-effective, high-performance, and attractive mobile device designs for the consumers. The carriers have responded to the increase in data rates by building a roadmap for the increasingly efficient use of the frequency bands that they have licensed. The mobile phone OEMs have improved the phone handset by creating slimmer phones using cosmetically attractive materials (such as metal backs, metal rings) and operating in the frequencies on the carriers' roadmap. But the mobile phone OEMs are being challenged by the multiple frequency bands now used (from 700 to 2700 MHz expanding to 600 to 3500 MHz in the next 5 years), the aggressive device designs, and the ever-present consumer price and quality expectations. In particular, these requirements present a significant issue for the antenna performance in mobile devices. The use of antenna RF (radio frequency) tuning is the disruptive technology that is being used to respond to the antenna challenge.

The history of the development of BST (barium strontium titanate) started with the discovery of ferroelectricity in barium titanate in the 1940s. In 1988, Paratek Microwave Inc., a private start-up company, was founded to exploit the unique properties of BST in tunable devices. The basic BST tuning and materials technology was funded under a DARPA contract for the FAME (frequency agile materials for electronics) program and the vision of Paratek was to commercialize the technology for military and commercial applications. Since 2006, the commercial focus of the company was to implement tunable and adaptive antenna impedance matching in mobile devices. In 2011, Paratek Microwave introduced their first generation of BST-based tuning solutions into the mass market, and since then, RF tuning using various technology options has become a mainstream solution in modern smartphones. The mass production and market support for BST-based tunable solutions has been led by ON Semiconductor and STMicroelectronics—two licensees of Paratek's IP for BST devices. Paratek Microwave was acquired by BlackBerry® in March 2012. Fundamental capacitor and materials work has been pursued at many universities around the world including significant efforts in the United States at Pennsylvania State University, North Carolina State University, and the University of California at Santa Barbara.

This chapter will detail the device characteristics of the BST-based tuning technology and some specific attributes and considerations in designing tuning solutions to be used in the tunable and adaptive RF front end of a modern mobile device.

3.2 BST TUNABLE CAPACITOR TECHNOLOGY AND STRUCTURE

While many "tunable" capacitors are actually "switched" capacitors, where the capacitance value is changed in discrete steps using RF switches, barium strontium titanate, or BST as it is commonly called, is a ceramic material with an unusual property—its dielectric constant is a function of the electric field in the material.[1,2] Capacitors using BST as the dielectric are truly tunable and have a capacitance that is dependent on the DC bias voltage applied to the capacitor. Tunable capacitors based on BST material have proven themselves in the production of antenna tuning solutions used in millions of mobile phones worldwide. BST tunable capacitors provide a change in capacitance or tuning ratio of 4:1 combined with low losses at microwave frequencies, high linearity, and up to 10 W power handling. The devices are produced as integrated circuits on 6 in. wafers using many of the process techniques used in the silicon IC industry and are available from multiple sources. As the need for better broadband antennas covering wider frequency ranges is driven by the design of attractive phones in small form factors, the use of antenna tuning will continue to increase and the BST tunable capacitors of today and tomorrow will be there to meet the challenges.

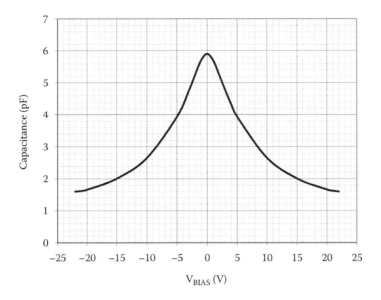

FIGURE 3.1   Capacitance–voltage relationship of BST capacitor.

The dependence of the capacitance on the bias voltage of a typical BST tunable capacitor is illustrated in Figure 3.1, where the capacitance in picofarads is on the vertical axis and applied voltage in volts is on the horizontal axis.

At zero applied voltage, the capacitance is 5.7 pF but at 22 V, in either polarity, the capacitance drops to 1.59 pF. Four to one changes in capacitance can be easily observed in such capacitors. So rather than fixed increments of capacitance change, the BST capacitor offers a continuous set of capacitance values, depending on the applied voltage, which lie between a maximum at 0 V and a minimum at some maximum applied voltage, which is usually determined by the breakdown voltage of the capacitor or the long term reliability of the device.

3.2.1 BARIUM STRONTIUM TITANATE

BST is a solid solution of two ternary oxides, barium titanate (BTO) and strontium titanate (STO). This ceramic material has two forms depending on a temperature defined as the Curie temperature. Below the Curie temperature, Tc, the Ba^{2+}/Sr^{2+} and Ti^{4+} cations in the unit cell have a tetragonal structure[3] with a net dipole moment and the material is ferroelectric. In the ferroelectric state, there is hysteresis of the electric field-polarization curves that results in losses making it unsuitable for tunable capacitor applications. Lowering the Curie temperature of the material below the desired operating temperature by changing the ratio of BTO to STO, the material structure is then cubic (Figure 3.2) and the material has no dipole moment. The electric field-polarization response is no longer ferroelectric with hysteresis but paraelectric with low losses and well suited to RF and microwave applications.

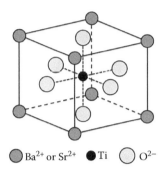

Ba^{2+} or Sr^{2+} ● Ti   ◯ O^{2-}

FIGURE 3.2 $Ba_xSr_{(1-x)}TiO_3$ unit cell in paraelectric state above Curie temperature.

The Curie temperature, Tc, of BST can be reduced from about 120°C for $BaTiO_3$ to about −45°C for a 50:50 mixture of BTO and STO, $Ba_{0.5}Sr_{0.5}TiO_3$. The relative ratio of BTO and STO also affects the tuning and loss characteristics of the capacitors. At high barium compositions, the tuning effects are maximized but the microwave loss of the material is high. At low barium levels, the loss is good but the tuning ratio is low. For most practical applications, the barium composition is between 35% and 55%, resulting in low-loss and good tuning while operating above the Curie temperature in the paraelectric state. A proprietary BST material composition known as ParaScan™, developed by Paratek Microwave, further enhances the performance of BST by adding small quantities of additional elements (dopants) to further lower the microwave loss and leakage currents.

3.2.2 Capacitor Structure

In its simplest form, the tunable capacitor is no more than a parallel plate capacitor using ParaScan or BST as the dielectric between the two plates as shown in Figure 3.3. A DC potential is applied between the top and bottom electrodes to establish the electric field in the ParaScan and therefore, set the capacitance value as seen in Figure 3.1.

Parallel-plate tunable capacitor

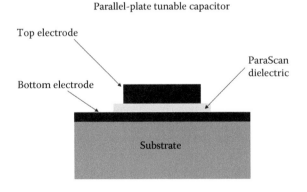

FIGURE 3.3   Tunable capacitor structure.

The thickness of the ParaScan layer sets the required tuning voltage as well as determines the capacitance per unit area (pF/mm^2) which drives the physical size of the capacitor.

The choice of the barium composition of the tunable material, the substrate material, the capacitor electrodes, and the processing used to make the capacitor largely determine the tuning ratio and leakage current of the capacitor. Current ParaScan tunable capacitors achieve 4:1 tuning ratios at applied bias voltages from 2 to 20 V. The leakage currents at 20 V are less than 1 nA. The conductivity and thickness of the metal layers determine the Q at the microwave frequencies. Of course in real life, there are additional design elements to consider such as a bias network, interconnect metals, dielectric passivation, and the need to address linearity and power handling for applications such as mobile handsets. These complicate the simple capacitor layout of Figure 3.3 and lead to an integrated circuit comprising multiple tunable capacitors and resistors on each die, which is referred to as a passive tunable integrated circuit (PTIC). For clarity, we will use *tunable capacitor* to refer to a single parallel plate capacitor structure and *PTIC* when discussing an integrated circuit with multiple capacitors. In addition, the term *stack* will be used to describe the number of capacitors connected in series in a PTIC.

3.3   BST DEVICE DESIGN WITH APPLICATION TO RF PERFORMANCE

The complete design of BST-based tunable capacitors for use in mobile phone and other applications must consider the DC and RF electrical design and the acoustic design, each of which will be discussed in the following sections. With a good electrical design, the capacitor can work well from 500 to 2700 MHz with low loss in antenna matching circuits. The capacitor must withstand the high peak RF voltages often present at the terminals of the antenna and perform without introducing significant distortion of the RF signal. Let's see how each of these requirements is met.

3.3.1   DC AND LOW-FREQUENCY CHARACTERISTICS

The performance characteristics of a typical BST-based tunable capacitor are illustrated in Table 3.1 for a 4.7 pF device.[4,5] The bias voltage for tuning is 2 V for the

TABLE 3.1
Typical Characteristics of a 4.7 pF PTIC

Characteristic (25°C)	Typical Value	Units
Operating bias voltage range	2–20 V	volts
Capacitance, 2 V	4.7	pF
Capacitance, 20 V	1.175	pF
Tuning range	4:1	
Leakage current, 20 V	<100	nA
Frequency range	500–2700	MHz

maximum capacitance, in this case 4.7 pF, and 20 V for the minimum, or 1.175 pF. The tuning ratio of the capacitor is, therefore, C(2)/C(20) or 4:1. Higher tuning ratios can be achieved at higher bias voltages but careful attention must be paid to the reliability of these devices. The leakage currently associated with the bias voltage at its maximum of 20 V is less than 100 nA and usually less than 10 nA. The power draw of the capacitor at the maximum leakage is only 0.1 μW. These devices cover frequencies from 500 to 2700 MHz and are specified over the 700 to 2700 MHz range typical of a modern mobile handset.

To meet the linearity and voltage handling requirements, the PTIC uses many capacitors connected in series on a single die hence the designation of this as an integrated circuit. A typical PTIC is shown in Figure 3.4.

This PTIC integrates 24 series connected capacitors, each of which is tunable to create the desired two terminal capacitance at the RF_{IN} and RF_{OUT} ports of the IC. The DC bias voltage is applied to the terminal marked V_{BIAS} and the RF_{IN} and RF_{OUT} terminals must be at DC ground. The bias voltage is distributed inside the IC through a network of 24 resistors supplying the DC bias to each capacitor while at the same time providing isolation of the DC and RF signals. The resistor values must be large enough to avoid the degradation of the Q due to resistive losses while staying small enough to permit fast charging and discharging of the capacitors as the DC tuning voltage is changed. In practice, the bias networks limit the Q of the PTIC up to 500 MHz, below the desired operating range, yet still support capacitor transition times of 100 microseconds and less.

While the use of multiple capacitors in series improves the voltage handling and linearity as will be discussed further, the total capacitance on the PTIC die is much greater than the net series capacitance in the RF circuit. If there are N individual capacitors in series to produce a net capacitance of C_{NET} between the RF_{IN} and RF_{OUT} ports of Figure 3.4, each capacitor must be $N \times C_{NET}$ and the total capacitance on the die is now $N^2 \times C_{NET}$. For example, a typical 4.7 pF PTIC with 24 individual

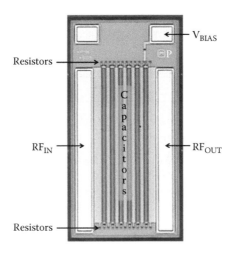

FIGURE 3.4   Photo of a PTIC with 24 series-connected capacitors and bias resistors.

capacitors in series must use a string of 112.8 pF capacitors for a total capacitance on the die of 2707 pF. Yet, the very high relative dielectric constant of the BST dielectric, between 400 and 450, keeps the size of the capacitors small so that the total die area including the resistor bias network and pads is below 1 mm^2.

3.3.2 QUALITY FACTOR

While a tunable circuit can optimize the impedance match of an antenna or amplifier to get the best response, the loss of the tuning elements may reduce any advantage. Therefore, the loss of the PTIC, expressed as the Q or Quality Factor, is a key factor and one where the BST-based tunable capacitor technology performs well. At low frequencies where the bias network loss dominates the PTIC response, the Q is determined by

$$Q = \omega R_p C_p \qquad (3.1)$$

where
  ω is 2π times the frequency in Hertz
  R_p is the parallel equivalent resistance of the bias network in ohms
  C_p is the parallel equivalent capacitance in farads

Note that the Q of a parallel RC circuit rises with frequency. At very low frequencies, the effective resistance of the leakage current of the capacitors could also be included but the very low leakage of PTICs, below 100 nA at 20 V, makes this a negligible factor.

The Q at high frequencies, where the series resistance of the capacitor plates and interconnects dominates is given by

$$Q = \frac{1}{\omega R_s C_s} \qquad (3.2)$$

where
  R_s is the series equivalent resistance in ohms
  C_s is the series equivalent capacitance in farads

The Q dominated by series losses falls with increasing frequency.

Before examining the measured Q of PTICs, some discussion of the measurements used to obtain accurate Q values is needed. For a high Q capacitor, the impedance is nearly all capacitive reactance with very little resistive loss. So the impedance on a Smith chart is located in the bottom half plane and out very near the edge of the chart. This is a very difficult region to get accurate data, especially for capacitors with Q values at or above 100. An approach that has worked well is to use coplanar RF probes along with custom calibration standards to extract the Q. The RF probes are designed to fit the spacing of the PTICs. While some specific test structures are used, the final measurement of the Q must be taken on the actual PTIC device and layout. By standardizing some of the device dimensions in the PTIC designs, only a few sets of these coplanar probes are needed for all the PTIC measurements.

TABLE 3.2
Parameters for a Simple Model of a 4.7 pF Tunable Capacitor

Parameter	Value	Comments
Capacitance	4.7 pF	
Parallel resistance	35 kohms	
Series resistance	0.1 ohms	
Material Q	120 at 1 GHz	Frequency dependence is $f^{-0.12}$ for frequency in GHz
Tuning ratio	4:1	

Considering first the Q response at 0 V, we have three contributions to the Q of the capacitor or PTIC: the bias network, the series resistance of the metals and interconnects, and the tunable material itself. The bias network appears as a parallel loss mechanism and impacts the Q at low frequencies, typically up to about 500 MHz, as in Equation 3.1. The series resistance losses impact the Q at higher frequencies above 2 GHz or so as defined in Equation 3.2. And the material losses extend over the whole frequency range. The values for the series and shunt resistances, material Q, and capacitances are shown in Table 3.2.

All three factors are graphed in Figure 3.5 for a 4.7 pF tunable capacitor at zero bias from 100 MHz to 3 GHz using a log scale for both the frequency and the Q.

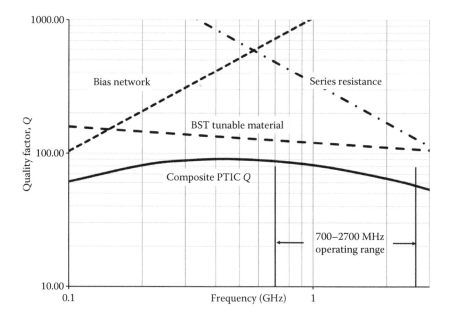

FIGURE 3.5 Total Q of the tunable capacitor at zero bias is described by the losses due to the bias network, the series resistance, and the BST material.

The Q due to the bias network resistance, labeled bias network, rises with frequency starting at about 100 at 100 MHz and reaching 1000 at about 1 GHz. The Q due to the series resistance falls from 1000 at about 340 MHz to a little above 100 at 3 GHz. And the material Q representing losses in the tunable BST material falls slowly from 158 at 100 MHz to 105 at 3 GHz. This slow drop in the material Q with frequency is described by

$$Q_m = Q_0 \times \left(\frac{f}{f_0} \right)^n \tag{3.3}$$

where Q_m is the material Q at a specific frequency f_0. The frequency dependence of the material, n, is generally near zero. Further, the distinction between a parallel and a series capacitance as indicated by C_s and C_p in Equations 3.1 and 3.2 has been ignored as they are equal when the Q is high.

The Q of the tunable capacitor incorporating all three loss mechanisms is shown by the line labeled *Composite PTIC Q* in Figure 3.5 and curving gracefully up from 61 at 100 MHz to peak at 90 at 440 MHz and then roll off to 54 at 3 GHz. The Q at the lower frequencies is dominated by the material losses and the bias network losses. At the upper frequencies, the Q is dependent on the material losses and the series resistance loss. In the frequency range from 500 MHz to 2.7 GHz, the capacitor Q ranges from 87 to 56, excellent values for tunable circuits.

The measured Q of 4.7 pF PTICs using 24 tunable capacitors in series at 2 V bias is illustrated in Figure 3.6.

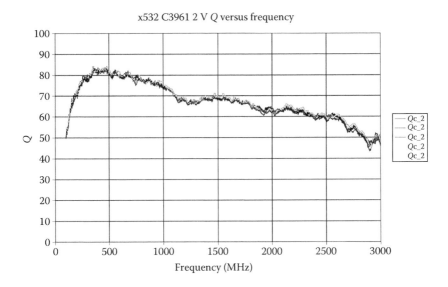

FIGURE 3.6 Measured Q at 2 V of five 4.7 pF PTICs, each with 24 tunable capacitors in series with bias networks.

These curves represent the measured responses of five devices across a wafer. The data is very similar to the curve we generated from a few simple equations in Figure 3.5. One difference is the use of linear scales now in frequency and Q in Figure 3.6 to better show the response in the 500–2700 MHz region where this device would be used. These devices were biased at 2 V for this data, which is the lowest recommended operating voltage for them. It may be worthwhile to point out that the devices being used to illustrate the performance of BST-based tunable capacitors are not test devices but complete production devices with multiple series capacitors, typically 24, and complete bias networks using a production process designed to create robust, high-performance PTICs providing reliable performance in mobile phones.

Looking now at the Q variation with frequency when the PTIC is biased to its lowest capacitance value, the simple Q relationships of Equations 3.1 and 3.2 still hold but, as we will see, the tunable material now introduces an acoustic response which must be included in the capacitor design. Looking first at the impact of the bias network, the series resistance, and the material losses, the bias network losses become more important than the series resistance losses. Remember that when the capacitor is biased at high voltage, the capacitance is reduced by about a factor of 4 in these capacitors. So the Q of the bias network, which varies proportionally to the capacitance, drops by a factor of 4 as shown in Figure 3.7.

Where the bias network Q at 100 MHz was 103, it is now 26. This pulls down the Q below 1 GHz but the chosen design values maintain a Q above 60 from 500 MHz to 1 GHz. Consider that the impedances in the bias network could be raised and the Q improved but the response time of the capacitor would then

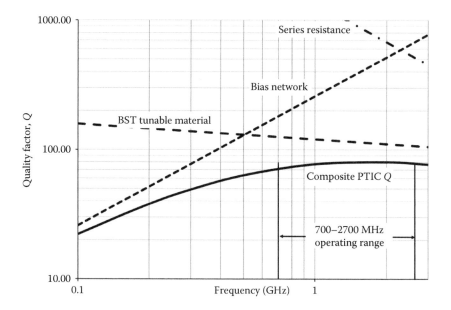

FIGURE 3.7   Total Q of the tunable capacitor at maximum bias as described by the losses due to the bias network, the series resistance, and the BST material.

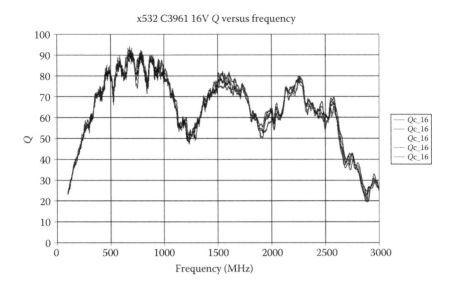

FIGURE 3.8   Measured Q at 16 V of five 4.7 pF PTICs, each with 24 tunable capacitors in series plus bias network.

be slowed. Since the Q due to the series resistances is inversely proportional to the capacitance, the series resistance losses have less effect at these high bias voltages and the composite Q of the capacitor remains high through 3 GHz.

When we look at the measured Q over frequency of Figure 3.8 for a capacitor biased at the maximum tuning voltage, we can see that the simple model for Q is no longer sufficient to describe the Q. Comparing Figures 3.6 and 3.8, the overall Q at 100 MHz has dropped from about 60 to about 23 as expected due to the parallel restive losses of the bias network, and the Q at the higher frequencies is good except for the appearance of resonances at about 1250 and 1900 MHz, which cannot be attributed to the causes we have been discussing. So there must be some new effect.

To understand what is happening to the Q response under bias, let us start again with the case of no bias or very low bias. When there is no electric field and the temperature is above the Curie temperature, the $Ba_xSr_{(1-x)}TiO_3$ unit cell of Figure 3.2 has a titanium atom in the center of a cube, a barium or strontium atom at each of the eight corners of the cube, and six oxygen atoms at the center of each face of the cube. This configuration has no dipole moment. But when an electric field is applied, the position of the atoms shift causing a large dipole moment or polarization of the crystal, which we exploit as a change in the permittivity of the BST, in other words, our tuning. But in addition to the change in dielectric constant, the material becomes electrostrictive, exhibiting a piezoelectric response to an extent determined by the strength of the applied electric field.[6-8] Therefore, the observed dips in Q are caused by bulk acoustic resonances excited by the piezoelectric response of the BST. This has practical implications for tunable capacitors as the observed resonances can destroy the Q at the desired operating frequencies. An acoustoelectric model of the

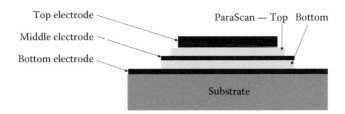

FIGURE 3.9   Cross section of two-layer capacitor structure used to reduce acoustic excitation of resonances.

capacitor must be used to determine the optimum layer thicknesses and composition of the materials in the capacitor stack to tune the acoustic response.

To further address the acoustic response, a two-layer capacitor design[9] can be used to reduce the acoustic excitation of the capacitor structure. As shown in the cross section of Figure 3.9, two capacitors are fabricated, one on top of the other, the top capacitor from the top electrode to the middle electrode and a bottom capacitor from the middle electrode to the bottom electrode. When an RF signal is applied across the two capacitors from top electrode to bottom electrode, the RF signals on the two capacitors are approximately equal and in phase. By biasing the middle electrode with the tuning voltage, typically a positive voltage from 2 to 20 V, and DC grounding the outer two electrodes, the DC electric field in the two capacitors is inverted relative to the RF signals. That is, in the upper capacitor, the RF signal flows from the grounded terminal to the positive terminal while in the lower capacitor the RF flows from the positive terminal to the grounded terminal. As a result, the acoustic excitation of the top capacitor is out of phase and cancels the acoustic excitation of the bottom capacitor. This two-layer configuration is like flipping a switch to reduce the acoustic coupling coefficient of the tunable material. One more result of the two-layer structure is the cancellation of the harmonic generation of even-order harmonics (2, 4, 6,...) as will be seen in our discussion on linearity.

3.3.3 Linearity

Due to the sophisticated modulation used in modern mobile communications systems, the linearity of any tuning device is very important to the performance of a mobile phone. This linearity is usually expressed as requirements on third-order intermodulation and harmonic generation. In the PTIC, the use of N multiple capacitors in series decreases the RF voltage across any individual capacitor by N to improve the linearity. In fact, a doubling of the number of capacitors in series decreases the voltage across each capacitor by a factor of 2 and the IP3 drops by 6 dB. For this reason, even higher stacks of capacitors than the current 24 are being developed. The third-order intermodulation point is measured using two closely spaced RF signals, F_L and F_H, and recording the level of the third-order intermodulation products, $F_1 = 2F_L - F_H$ and $F_2 = 2F_H - F_L$. Measuring F_1 and F_2 in a small signal range for the device under test, the intermodulation rises at three times the rate of the fundamental tones, F_L and F_H. The intercept point is then found from the extrapolation of the two curves, one for the fundamental tones and the other for the intermodulation products. This intercept is the third order intercept point or IP3 for short.

The IP3 of PTICs has been characterized by mounting the PTIC under test in shunt across a 50 ohm line and also by mounting it in series in the 50 ohm line. Extensive filtering of the two test signals must be used to ensure that the signals are clean. Attenuators are used to provide isolation between the two source signals and prevent reflections from the filters out of band. The measured shunt IP3 of three 4.7 pF PTICs is shown in Figure 3.10. The IP3 in each case is plotted against the applied DC tuning voltage. The responses of the three PTICs are nearly identical. The measurement frequencies, F_L and F_H, were 824 and 846 MHz at power levels up to 25 dBm. The shunt configured 4.7 pF PTIC exceeds 69 dBm IP3 from 2 to 20 V and over 75 dBm starting at 4 V.

Any nonlinear device will generate harmonics when excited by an AC signal. The PTIC design using multiple tunable capacitors connected in series works to reduce the harmonic generation by decreasing the RF voltage across any one capacitor. The harmonics are tested by placing the PTIC in shunt with a 50 ohm line and filtering out the fundamental signal before measuring the harmonics on a spectrum analyzer. Attenuators are again used to eliminate the effects of mismatches in the test setup. The measured harmonics of two 4.7 pF PTICs at 824 MHz in shunt are shown in Figure 3.11.

The second harmonics are −60 dBm and lower over the specified operating range of 2–20 V, in part because the two-layer capacitor structure suppresses the generation

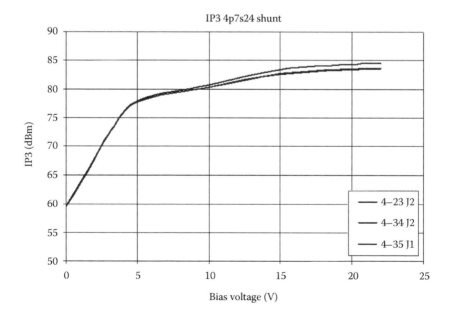

FIGURE 3.10 IP3 measurements of a 4.7 pF PTICs in shunt.

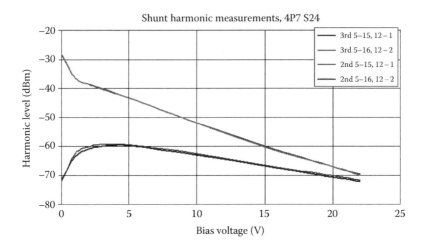

FIGURE 3.11   Second and third harmonics of a 4.7 pF PTICs in shunt at 824 MHz.

of all even harmonics. The third harmonic stays at or below −38 dBm from 2 to 20 V bottoming below −65 dBm at 20 V.

3.3.4   POWER HANDLING

The tunable matching circuits in the transmit paths of mobile phones must be able to handle several watts of RF power at times. Although the RF loss of the PTIC is low, at these power levels, significant RF heating can occur and the tunable devices must be designed to dissipate the heat. Also, the breakdown voltage of the tunable capacitors is reduced as their temperature rises, so a safety margin is needed. The design of the PTIC addresses this in two ways: through optimizing the structure of the capacitor to achieve high breakdown voltages and through the spreading of the heat.

The BST or ParaScan tunable material, the capacitor electrodes, the associated dielectric layers, and metal interconnects, all must be chosen and optimized to achieve high breakdown voltages. Figure 3.12 shows Weibull distribution plots of the breakdown voltage taken at room temperature of over 40 devices each from 2 wafers populated with PTICs.

Notice the tight distribution of the breakdown values indicating excellent uniformity across the wafers. The median breakdown is about 119 V, sufficiently high to permit operation at high power even at +85°C.

The design choice to use 24 series connected tunable capacitors in the PTIC also contributes to the power capability by reducing the applied voltage of the RF signal on individual capacitors by a factor of 24. But particular attention must be paid to the design of the resistors in the bias networks to be able to withstand the RF voltages at high power. Using alumina or sapphire substrates for the PTIC process, the 24 series connected capacitors also spread the heat across the PTIC die and reduce the temperature rise.

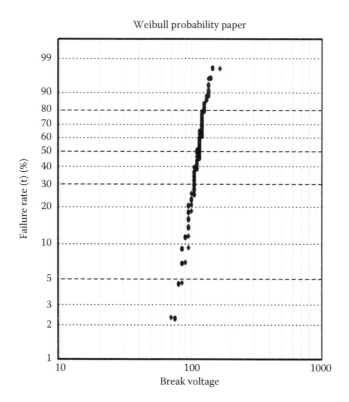

FIGURE 3.12 Breakdown voltage distributions across two wafers of PTICs, 40 devices per wafer.

The final test for the power handling is conducted using a high-power RF source, typically a 10 or 20 W TWT amplifier, and a PTIC mounted in series in a 50 ohm line terminated with a 50 ohm load. Attenuators or isolators placed between the TWT and the device under test protect the TWT from the high mismatch should the PTIC fail. With the high breakdown voltage and series connected design of the PTIC, power levels of 10 W are routinely achieved without failure.

3.4 RELIABILITY AND QUALIFICATION

PTICs meet all the lifetime and qualification requirements for reliable operation in mobile applications such as handsets. The recommended qualification tests for operating life (high-temperature operating life or HTOL) are based on the characteristic reliability factors of the BST (barium strontium titanate) tunable dielectric and the mission life of the phone.

3.4.1 ACCELERATION FACTORS

The reliability of ceramic capacitors has been well characterized for electronic and medical applications.[10,11] Under high DC bias voltages and high temperatures,

BST capacitors generally fail due to (1) thermal runaway with a gradual increase in leakage current which eventually shorts the capacitor, (2) an abrupt increase in current, an avalanche, shorting the capacitor, or (3) a mixture of both failure modes. The PTICs made with ParaScan dielectric characteristically fail due to an avalanche short in one of the multiple series capacitors on the chip leading to a high leakage current, limited by the series resistance of the on-chip bias network. While 1 of the 24 capacitors in the PTIC has failed, the other 23 continue to operate normally and the device still operates, though with reduced tuning range and higher current draw. For many applications, this "soft" failure mode provides continued operation of the tuned product with reduced performance but without a failure of the final product containing the PTIC.

The failure of a single capacitor in the PTICs is accelerated by both temperature and voltage factors and is modeled following the work of T. Prokopowicz and A. Vaskas[12] as

$$MTTF = A \times V^N \times e^{E_a/kT} \tag{3.4}$$

where

E_a is the thermal activation energy, eV
T is the temperature, degrees Kelvin
k is Boltzmann's constant
V is the bias voltage, V
N is the voltage acceleration factor
A is an arbitrary constant

Based on measured data generated over multiple wafer lots at voltages up to 24 V and temperatures to 150°C, Paratek devices using ParaScan dielectrics typically exhibit the following acceleration factors.

$$E_a = 1.01 \text{ eV} \quad \text{and} \quad N = -8.3 \tag{3.5}$$

These two factors allow the PTIC performance over time to be understood and verified for the intended application.

3.4.2 MISSION LIFE PROFILE

To understand the relation of the measured lifetime to the lifetime of the PTICs in a phone, a mission life profile based on time and temperature is constructed to represent the environment in the phone. Using a profile based on a 55°C board temperature and voltages up to the maximum 20 V operating voltage, equivalent test hours at 125°C and 20 V are calculated.

- Out of each 24 hours, the phone is both operating (PTICs biased) and non-operating (PTICs unbiased).

- When the phone is operating, the temperature of the PTICs in the phone is higher than the circuit board temperature due to heating by other components surrounding the PTIC components such as the RF power amp. Multiple values for the temperature rise are based on different operation modes of the phone, such as hotter for high-power GSM and cooler in less power hungry modes.

Note that the mission profile represents aggressive use of the phone; 4 hours of use per day for a 30 day month equates to 7200 minutes per month use, more than most users consume on their phone plan.

3.4.3 EARLY FAILURE RATE

The target early failure rate for PTIC products at 2 years operating life is less than 200 ppm defects. There are several ways to measure this.

1. Measure the MTTF of 50 or more devices, plot the results on a Weibull plot and extrapolate the resulting curve to the 2-year mission life point. One problem with this approach is that the first failure out of 50–100 parts tested lies at only the 1%–2% failure level, but the extrapolation must extend by two orders of magnitude to the 0.02% point and is subject to any uncertainty in the slope of the failure rate at such low levels.
2. Measure 4500 devices to the 2-year mission life point and show zero failures to project less than 200 ppm at 60% confidence level. This approach more directly measures the failures at the 2-year mission life point, but requires many more devices to be tested and usually covers a small sample of wafers or wafer lots.
3. Institute a regular sampling of 10–100 devices from every production wafer to build a significant database of the failure points to allow calculation of the failure rate at the 2-year mission life point with greater accuracy.

A study conducted by Paratek Microwave followed this third approach performing a life test on 40 devices from every production wafer over an extended time frame to amass data on over 25,000 devices. The devices used in this study came from 644 production wafers fabricated in the ON semiconductor facility in Burlington, Ontario. A total of 25,760 devices were tested, 40 per wafer, with data at the 2-year, 5-year, 9-year, and 30-year life points. The data is summarized in Table 3.3.

TABLE 3.3
Early Failure Rate Data, 25,760 PTICs

644 Wafers/25,760 Devices	2-Year Life	5-Year Life	9-Year Life	30-Year Life
# Failures @	2	16	33	129
Failure rate (ppm)	78	621	1281	5008
Failure rate (ppm @ 60% CL)	121	689	1356	5146

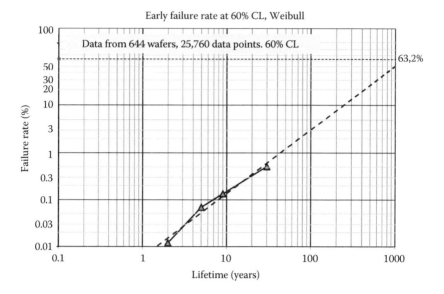

FIGURE 3.13   Weibull graph of life test of 25,760 PTIC devices to determine failure rate at 2-year handset operation.

As can be seen, there were 2 failures out of 25,760 devices at the 2-year mission life point or a 78 ppm failure rate. Applying a 60% Confidence Level to the data, the failure rate is only 121 ppm, nicely below the 200 ppm target. The data is graphed in the Weibull plot in Figure 3.13.

From this data, we see that not only are the BST-based PTICs better than the 200 ppm failure rate target at the 2-year mission life point, but the extrapolation of the data to MTTF suggests over 1000 years mean time to failure in the phone application.

3.4.4   QUALIFICATION TESTS

While reliable operation with low failure rates is required for PTIC products to be used in mobile handsets, there are other important qualification tests as well. These are summarized in Table 3.4. PTIC devices have passed all these qualification tests.

3.5   TUNING APPLICATIONS

While the design of an antenna-matching network for a mobile phone is not within the scope of this chapter, it is worthwhile to discuss some of the approaches used in working with the PTICs and then look at the commercial introduction of BST-based tuners in mobile handsets.

3.5.1   GENERIC TUNER

A generic tuning circuit is shown in Figure 3.14. This is a Pi circuit using three PTICs with two in shunt and one in series. A T circuit with two PTICs in series

TABLE 3.4

Qualification Test Results for PTIC Devices

Qualification Test	Test Method	Conditions	Qty Per Lot/#Lots	
Temp cycle (TC)	JEDEC JESD 22-A104	−40°C to +125°C 1000 cycles (Cond G)	45/3	PASS
High-temp storage (HTS)	JEDEC JESD 22-A103C	+150°C for 1000 hours	45/1	PASS
Low-temp storage (LTS)	JEDEC JESD 22-A119	−40°C for 168 hours	45/1	PASS
Drop	JEDEC JESD 22-A119	1500 G Peak 0.5 ms Pulse 3 Die Level 3 PWB Mounted	6/3	PASS
Vibration	MIL-Std-883 MTD 2007	Cond A 20G peak, 20–2000 Hz	6/3	PASS
Solderability	JEDEC JESD 22-B102E	Method 1	3/3	PASS
ESD	JEDEC JESD 22A114	Human Body Model	3/3	Class 1A

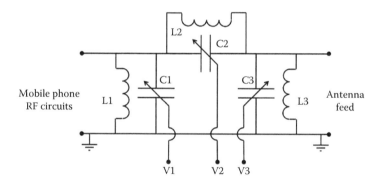

FIGURE 3.14 Generic tuner schematic using three PTICs.

and one PTIC in shunt might equally be used instead. The purpose of the tuner is to transform the antenna impedance to 50 ohms of the RF circuits in the mobile phone. The PTICs are C1, C2, and C3 which along with their control voltages V1, V2, and V3, provide variable capacitance values to alter the impedance transformation between the mobile phone's RF circuits on the left and the antenna feed impedance on the right. The degree to which the impedance transformation can be changed is related to the tuning range of the PTIC and the number of PTICs in the tuner. While this example shows three PTICs, many tuners require only two, or in some cases, one PTIC to achieve the needed response. Likewise, the three inductors shown may not all be required either. However, the two inductors, L1 and L3, on the two RF ports of the tuner provide the DC path to ground needed to keep the RF_{IN} and RF_{OUT} ports of the PTICs at 0 V. They also provide protection from ESD damage, particularly at the antenna feed where the typical phone spec of 6000 V or more greatly exceeds the capability of the PTIC itself.

In an open-loop tuner, the required values of the bias voltages are determined during the design of the tuner for that phone model and that particular antenna and are stored in memory on the phone. Knowing the operating mode (full or half duplex) and the frequency of the signal, the proper voltages can be looked up in memory and applied. This open-loop operation provides a significant advantage for the antenna match, but it does not correct for the dynamics in the use of a phone such as the shifts in antenna impedance when the phone is in a pocket, on a table, held in the hand, or placed at the ear.

Closed-loop operation of the tuner can use a directional coupler to monitor the incident and reflected power from the antenna directly during use. An algorithm to perform corrections is incorporated in the software and the phone shifts the tuner voltages to continually optimize the tuner response. This method provides the best performance at the expense of a very small additional processor load.

The operation of PTICs in practice requires voltages up to 20 V that are not readily available in a mobile phone. For this reason, the manufacturers of the PTICs provide several silicon ICs designed to interface the PTICs to both an existing voltage source in the phone and to the digital signals needed to allow the phone to set the PTIC voltages, whether in open-loop or closed-loop operation. These are referred to as HVDACs, short for high-voltage digital to analog converters.[13,14] Running from the 1.8 V supply in the phone, the HVDAC generates a 22 V or higher internal voltage. Available with either an SPI interface or a MIPI RFFE interface and from two to six outputs to drive up to six PTICs, the HVDAC uses an individual DAC for each output to convert the digital input to the desired output voltage for each PTIC.

3.5.2 COMMERCIAL ANTENNA TUNERS

In 2010, the first commercial implementation of this technology was realized in the Samsung Galaxy S II phone targeted for the Australian and Japanese markets. A picture of the implementation is seen in Figure 3.15 where the tuning circuit including the passive components as well as the PTIC tuning elements was housed in a RAFT™ (RF Adaptive Frequency Tuner) module.

This tuner product format was also utilized by HTC in both their Vivid Smartphone and in their Jetstream tablet. In February 2013, HTC introduced their flagship HTC One models which continued to use Paratek-based technology but for the first time implemented a discrete tuning solution using PTICs in plastic QFN packages as seen in Figure 3.16.

In March 2012, Paratek Microwave, under the leadership of CEO Ralph Pini, was acquired by Research In Motion (RIM, currently known as BlackBerry Inc.). BlackBerry aggressively implemented the PTIC antenna tuning technology in a number of phones that used their new BB 10 operating system. These BlackBerry phones all adopted a discrete tuning solution using a wafer level chip scale package

FIGURE 3.15 Tuning module including PTIC elements used in Samsung Galaxy S II mobile phone released in April 2011.

FIGURE 3.16 Discrete tuning solution used in HTC One models with PTICs in plastic QFN.

FIGURE 3.17 Antenna tuning section of BlackBerry Z10 with discrete wafer level chip scale packaged PTICs.

(WLCSP) for the PTIC, Figure 3.17, which created the smallest and lowest cost tuner implementation. In March 2013, the Z10 was introduced to the market, followed by the Q10 and Q5 phones.

In September 2013, BlackBerry continued to drive the Paratek BST technology with one of the first ever, closed-loop implementations of adaptive antenna tuning technology in their Z30 model. This was also the first implementation where both the main and diversity antennas in a single phone were tuned—a trend that is continuing to grow. In their press release of September 18, 2013, BlackBerry specifically referred to the antenna tuning as a key feature:

New Antenna Technology—The BlackBerry Z30 smartphone features BlackBerry's new generation antenna technology that dynamically tunes reception to give you better connectivity in low signal areas. BlackBerry® Paratek Antenna can give you faster data transfers and fewer dropped calls in low signal areas, keeping you connected in more places.

It was a big year in 2014 for the introduction of PTIC-based antenna tuning. A series of China-based OEMs started using the Paratek BST tuning technology—ZTE in their Nubia 5S and Huawei in their Asend P7 smartphones. In September 2014, BlackBerry announced its second phone using closed-loop tuning technology—the *Passport*. Also in September 2014, Motorola introduced the first of three phones with Paratek BST-based tuning technology—the Moto-X phone which was followed by the Droid Turbo and the Nexus 6. The introduction of these phones brought attention specifically to the improved antenna performance with what Motorola referred to as "dynamic tuning."

"…you *definitely* want it,…improvements in power to the antenna… of over 500%…" Phonearena.com

"…even better than the QFE15xx antenna tuner that Qualcomm has made…" Anandtech.com

"Sometimes, you can tell that a company went above and beyond in an effort to add innovative features to a smartphone that aren't just different for the sake of being different, they truly add value to the user experience…" BGR.com

At the 2015 Mobile World Congress held annually in Barcelona (Spain)—new phones were announced where Paratek-based BST tuning was used for improved antenna performance.

3.6   BST TECHNOLOGY'S FUTURE

The continued development of BST tunable technology will not only improve the performance of PTICs for antenna tuning applications but also expand the BST technology to other tunable functions. The roadmap for this technology in antenna tuning applications calls for larger tuning ratios, higher IP3, and better Q (lower loss). This will continue to make tunable capacitors based on BST material the leading choice for the mobile handset market. But, in addition, the extension of the BST capacitor to higher frequencies, up to 5 GHz, and lower operating voltages, down to 10 V or even 5 V for some applications, will open new opportunities.

The technology roadmap for the production of PTICs for tunable antenna solutions is focused on three key parameters: tuning ratio, linearity, and Quality factor (Q). The goals for the next generation of PTICs have been summarized in Table 3.5, along with the current performance for each parameter.

The tuning ratio will increase from the current 4:1 ratio at 2–20 V bias to a new ratio of 5:1. This is being accomplished through optimization of the tuning material and the electrode interfaces and other material used for the capacitors. Additional

TABLE 3.5

Improvements to PTIC Performance on a Near-Term Roadmap

Parameter	Current Performance	Future Performance	
Tuning ratio	4:1	5:1	
Quality factor (Q)	60	90	500–1000 MHz
	50	70	1700–2700 MHz
Linearity, IP3	>68 dBm	>75 dBm	Over entire tuning range

improvement may also come from extension of the maximum operating voltage but staying within the voltage range of the existing HVDAC integrated circuits designed to provide the tuning voltage for PTICs. The increased tuning range will extend the area of impedances which can be tuned by matching circuits employing the PTICs and may, in some cases, allow a decrease in the number of PTICs needed for a specific application, say from three PTICs to two or from two PTICs to one.

The higher Q of these next-generation PTICs will lower the loss of tuning networks, particularly where a high standing wave ratio must be matched and the impedance transformations are great. These improvements rely on a lower loss ParaScan formulation for BST to help across the entire frequency range coupled with lower bias network losses and good acoustic design.

Finally, higher linearity means an increase in the stacking of capacitors in series beyond the typical 24 series connected capacitors as used now. As discussed earlier, for a PTIC of N series connected capacitors, the total capacitance on the die is N^2 times the desired net series capacitance. For a 4.7 pF PTIC with 24 capacitors stacked in series, the total capacitance on the die is therefore 2707 pF. If the stack were increased to 48, the IP3 will be 6 dB lower, but the die will now have four times the total capacitance or 10,829 pF! Designs with increased stacking have been shown to improve the IP3 and are now shipping in high volume, production handsets.

3.7   CONCLUSIONS

Tunable capacitors based on BST material have proven themselves in the production of antenna tuning solutions used in millions of mobile phones worldwide. BST tunable capacitors provide a change in capacitance or tuning ratio of 4:1 combined with low losses at microwave frequencies, high linearity, and up to 10 W power handling. The devices are produced as integrated circuits on 6 in. wafers using many of the process techniques used in the silicon IC industry and are available from multiple sources. As the need for better broadband antennas covering wider frequency ranges is driven by the design of attractive phones in small form factors, the use of antenna tuning will continue to increase and the PTICs of today and tomorrow will be there to meet the challenges.

ACKNOWLEDGMENTS

The development of ParaScan, the improved BST dielectric, the design of a robust tunable capacitor, and the qualification and production of PTICs are the result of the efforts of many people at Paratek Microwave, now part of BlackBerry Inc., along with ON Semiconductor in Burlington, Ontario, Canada, and STMicroelectronics in Tours, France. To all those involved, many thanks for the hard work bringing this technology to market. The contributions of Dr. James V. DiLorenzo, Mr. Ralph Pini and Dr. Louise Sengupta were critical to the successful development of PTIC technology and its application in commercial mobile phones.

The authors gratefully acknowledge the support of the BlackBerry Technology Services division of BlackBerry Inc. and ON Semiconductor Corporation in the publication of this information on BST technology and products. Figures 3.1, 3.4, 3.6, 3.8, and 3.10 through 3.13 as well as the data in Tables 3.3 and 3.4 are used with the permission of BlackBerry Inc. The data of Table 3.1 on a typical 4.7 pF capacitor are a combination of data from the data sheets of ON Semiconductor and STMicroelectronics.

REFERENCES

1. Tagantsev, A.K., Sherman, V.O., Astafiev, K.F., Venkatesh, J., and Setter, N., Ferroelectric materials for microwave tunable applications, *Journal of Electroceramics*, 11, 5–66, 2003.
2. Newnham, R.E. and Cross, L.E., Ferroelectricity: The foundation of a field from form to function, *MRS Bulletin*, 30, 845–848, November 2005.
3. Remmel, T., Gregory, R., and Baumert, B., Characterization of barium strontium titanate films using XRD, *JCPDS—International Centre for Diffraction Data*, 38–45, 1999.
4. ON Semiconductor Datasheet, TCP-3147H Rev 0, 4.7 pF passive tunable integrated circuits (PTIC), http://onsemi.com, accessed September, 2014.
5. STMicroelectronics Datasheet, STPTIC, ParaScan™ tunable integrated capacitor, DOCID023772 Rev 3, http://www.st.com, accessed January 2014.
6. Gevorgian, S. and Vorobiev, A., DC field and temperature dependent acoustic resonances in parallel plate capacitors based on $SrTiO_3$ and $Ba_{0.25}Sr_{0.75}TiO_3$ films: Experiment and modeling, *Journal of Applied Physics*, 99(124112), 1–11, 2006.
7. Vendik, O.G. and Rogachev, A.N., Electrostriction mechanism of microwave losses in a ferroelectric film and experimental confirmation, *Technical Physics Letters*, 25(9), 702–704, 1999.
8. Tappe, S., Bottger, U., and Waser, R., Electrostrictive resonances in $(Ba_{0.7}Sr_{0.3})TiO_3$ thin films at microwave frequencies, *Applied Physics Letters*, 85(4), 624–626, 2004.
9. Oakes, J., Martin, J., Kozyrev, A., and Prudan, A., Capacitors adapted for acoustic resonance cancellation, U.S. Patent 7,936,553 (May 3, 2011).
10. Rawal, B.S. and Chan, N.H., Conduction and failure mechanisms in barium titanate based ceramics under highly accelerated conditions, Technical Information Publication, AVX Corporation, Myrtle Beach, SC, S-CFMB00M301-R, 1984.
11. Ashburn, T. and Skamser, D., Highly accelerated testing of capacitors for medical applications, *SMTA Medical Electronics Symposium*, Anaheim, CA, 2008.

12. Prokopowicz, T. and Vaskas, A., Research and development, intrinsic reliability, subminiature ceramic capacitors, Final report ECOM-90705-F, NTIS AD-864068, 1969.

13. ON Semiconductor Datasheet, TCC-106 Rev 4, Six Output PTIC Control IC, http://onsemi.com, accessed June 2014.

14. STMicroelectronics Datasheet, STHVDAC-304M, High Voltage BST capacitance controller, Doc ID 023054 Rev 2, http://www.st.com, accessed November 2012.

4 Tuned Antennas for Embedded Applications

Frank Caimi

CONTENTS

4.1   INTRODUCTION AND OVERVIEW

Antenna tuning is taking on greater significance as the number of frequency bands expands to meet the demands of users and network operators for higher data rates and greater network capacity. High-end mobile devices that can roam worldwide must also be able to operate seamlessly when traveling from country to country. To meet these demands the antenna must support numerous frequency bands even though operation in the past has been generally restricted to a single band or set of frequencies. With the evolution of LTE-Advanced this may change as simultaneous multiband operation (called inter-band carrier aggregation within the 3GPP standard) is being planned to allow for additional capacity expansion. An example of the number of bands currently being considered by 3GPP is shown in Table 4.1.

Not yet included in this list is expansion to lower frequencies, such as the 600 MHz band plan being considered by the Federal Communications Commission (FCC). The antenna, therefore, must be able to support each of these bands at least one at a time, so some consideration of the antenna design and radiation mechanism is necessary to understand how antenna tuning might be implemented to advantage.

4.2   RADIATION THEORY: BASICS, LOSS MECHANISMS, BANDWIDTH, AND RADIATION MECHANISMS

It is well known among antenna engineers that the design of any small antenna will be a trade-off between its dimensions and its electrical performance, and furthermore that physical laws determine the ultimate limitations inherent in any design. The physical limits imposed on any design can be simply stated: "bandwidth and antenna size are closely related," and "antenna gain and size are directly related." This means that the maximal gain of an antenna can be enhanced somewhat by varying the geometry and that the bandwidth can similarly be increased. A third degree of dimensionality is also afforded by the antenna designer: The efficiency of the antenna can be purposely degraded to additionally increase the bandwidth, but it will reduce the gain.

The relationship that describes the ultimate size versus bandwidth capability of an antenna was developed in several seminal papers by Harold Wheeler[1] and L.J. Chu,[2] and later by Roger Harrington.[3]

In 1946, Wheeler introduced the concept of the "radiansphere" and the volume relation to the maximum power factor achievable by the antenna. He related the energy within the radiansphere and outside the antenna, volume to the fundamental limitation of the antenna power factor. The power factor, it is reasoned, is proportional to the antenna volume V and also a shape factor. The nominal bandwidth BW

TABLE 4.1

LTE Frequency Bands and the Corresponding Regions

LTE Bands	Uplink (MHz)	Downlink (MHz)	Duplex Spacing (MHz)	BW (MHz)	Duplex Mode	Deployment in the World
Band 1	1920–1980	2110–2170	190	60	FDD	China, Japan, EU, Asia, Australia
Band 2	1850–1910	1930–1990	80	60	FDD	North and South America
Band 3	1710–1785	1805–1880	95	75	FDD	EU, China, Asia, Australia, Africa
Band 4	1710–1755	2110–2155	400	45	FDD	North and South America
Band 5	824–849	869–894	45	25	FDD	North and South America, Australia, Asia, Africa
Band 6	830–840	875–885	45	10	FDD	Japan
Band 7	2500–2570	2620–2690	120	70	FDD	EU, South America, Asia, Africa, Australia
Band 8	880–915	925–960	45	35	FDD	EU, South America, Asia, Africa, Australia
Band 9	1749.9–1784.9	1844.9–1879.9	95	35	FDD	Japan
Band 10	1710–1770	2110–2170	400	60	FDD	North and South America
Band 11	1427.9–1447.9	1475.9–1495.9	48	35	FDD	Japan
Band 12	698–716	728–746	30	18	FDD	North America
Band 13	777–787	746–756	31	10	FDD	North America
Band 14	788–798	758–768	30	10	FDD	North America
Band 17	704–716	734–746	30	12	FDD	North America
Band 18	815–830	860–875	45	15	FDD	North and South America, Australia, Asia, Africa
Band 19	830–845	875–890	45	15	FDD	Africa
Band 20	832–862	791–821	41	30	FDD	EU
Band 21	1447.9–1462.9	1495.9–1510.9	48	15	FDD	Japan
Band 22	3410–3500	3510–3600	100	90	FDD	

(Continued)

TABLE 4.1 (*Continued*)
LTE Frequency Bands and the Corresponding Regions

LTE Bands	Uplink (MHz)	Downlink (MHz)	Duplex Spacing (MHz)	BW (MHz)	Duplex Mode	Deployment in the World
Band 24	1626.5–1660.5	1525–1559	101.5	34	FDD	
Band 33	1900–1920		NA	20	TDD	
Band 34	2010–2025		NA	15	TDD	China
Band 35	1850–1910		NA	60	TDD	
Band 36	1930–1990		NA	60	TDD	
Band 37	1910–1930		NA	20	TDD	
Band 38	2570–2620		NA	50	TDD	EU
Band 39	1880–1920		NA	40	TDD	China
Band 40	2300–2400		NA	100	TDD	China, Asia
Band 41	2496–2690		NA	194	TDD	
Band 42	3400–3600		NA	200	TDD	
Band 43	3600–3800		NA	200	TDD	

is given as the power factor p multiplied by the resonance frequency f_0 according to the fundamental relationship:

$$V \propto p \propto \frac{BW}{f_0} \qquad (4.1)$$

In 1948, Chu extended Wheeler's analysis and expressed the fields for an omnidirectional antenna in terms of spherical wave functions and found limits for the antenna Quality factor Q, the maximum gain G_m and the ratio G/Q. Later in 1959, Harrington related the effects of antenna size, Quality factor, and gain for the near and far field diffraction zones for linearly and circularly polarized waves, and also treated the case where the antenna efficiency is less than 100%.

Additional work carried out by others[4–6] from 1969 through 2001 was directed to obtain exact expressions for the antenna Q over an expanded size range. These efforts led to the equation:

$$Q = \frac{1}{(ka)^3} + \frac{1}{ka} \qquad (4.2)$$

where
   k is $2\pi/\lambda$
   a is the radius of a sphere containing the currents associated with the fundamental or lowest mode, generally confirming the results obtained by Chu, Harrington, Collin and Rothschild, and others

As an antenna becomes smaller in relation to the wavelength, the stored energy increases, representing an increase in the reactive portion of the antenna impedance (reactive elements such as inductors and capacitors store energy). Antenna Q can be loosely considered as the ratio of the reactive term of the impedance to the resistive or dissipative portion.

$$Q \sim \left(\frac{\text{reactance}}{\text{resistance}} \right) \sim \frac{f}{\Delta f} \tag{4.3}$$

A higher Q implies a narrower bandwidth. So as the antenna becomes smaller the reactive portion increases in relation to the resistive portion, increasing the Q and reducing the bandwidth, readily observable from the Chu–Harrington curve. The best antennas are near the 100% efficiency Chu–Harrington curve indicating that the reactance to resistance ratio (i.e., Q) is as low as it can possibly be. That means that the antenna exhibits the broadest instantaneous bandwidth that it can have, without some additional tuning mechanism. Figure 4.1 shows the fundamental limit with various antenna types for illustration.

The resistance R is also an important factor in determining the antenna performance. The antenna resistance is composed of two dissipative portions—the EM

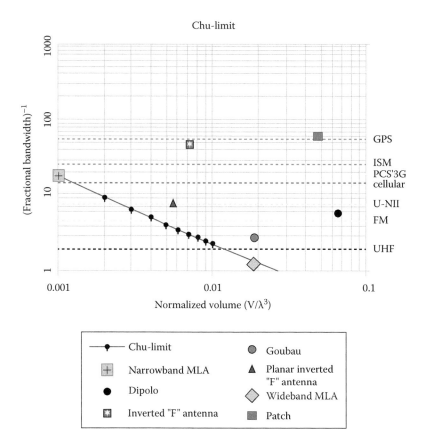

FIGURE 4.1 Chu–Harrington curve and antenna example Q relative to the normalized volume.

wave radiation into space and the loss due to heating in the conductors making up the antenna, the so-called "I-squared R" loss. To achieve efficient radiation of energy into space with minimum heating of the antenna elements, the resistive loss in the antenna conductors must be minimal with respect to the radiation resistance (loss). It turns out that both losses can be controlled somewhat through engineering design. As a result, poorly designed antennas will exhibit higher radiation than others for the same input power. A measure of the design quality is the *radiation efficiency*. Small antennas can have efficiencies that range from small fractions of 1% for submarine antennas to near 100% for others.

4.2.1 DESIGN PRINCIPLES AFFECTING SMALL ANTENNA DESIGN

4.2.1.1 Conductor Area

An antenna can be considered a lossy transmission line. A ½-wave dipole generally has an open circuit at one end and a connection to a low-impedance line at the other end. The same is true in the case of a ¼-wave monopole over a ground plane. The open end is a high-impedance point where the boundary conditions require that the current is zero. The forward wave propagates down the radiating element, reaches the end, and is reflected back to the source producing a reflected wave. The magnitude of the current at different portions along the line is determined by the boundary conditions and the sum of the reflected and forward propagating waves. The voltage along the line is also affected by the boundary conditions. The net current and voltage along the line, therefore, vary as a function of position from the end. If the line is nonradiating and non-absorptive, we say the line is lossless. The voltage and current vary in phase and the current at the end is zero where the voltage is maximum. Further along the line, there is a point that is reached where the current is maximum and the voltage is at a minimum. The distance from the end where the current increases from zero to a maximum is ¼ wavelength in a lossless transmission line. Antennas by the very nature must have loss to radiation, however, and depart somewhat from the ideal transmission line.

With this behavior, the loss resistance of the conductors becomes important in the antenna design. Since high-frequency currents flow at the outside of conductors and are governed by the skin effect, the effective resistance is related to excitation frequency and the conductor surface area. The basis for a good design is that conductors have a sufficiently large surface area to maintain low I^2R loss at points where the current is maximum. The points where the current is zero, or smaller, matter less and can have a smaller surface area.

4.2.1.2 Radiation Resistance

The radiation resistance of antenna depends on its geometry, current distribution along the radiating elements, and wavelength. For a short dipole, the resistance is given approximately by

$$Rrad_{dipole} = 200\left(\frac{L}{\lambda}\right)^2 \tag{4.4}$$

where L is the dipole length.

For a loop antenna, the radiation resistance can be larger:

$$Rrad_{loop} = 300,000 \left(\frac{b^2}{\lambda^2} \right)^2 \qquad (4.5)$$

where b is the loop radius.

The loop can have a higher radiation resistance at higher frequency as compared to a short dipole. An example calculation gives $Rrad_{dipole} = 5$ ohms at 1900 MHz for a length of 1 in., and $Rrad_{loop} = 12$ ohms for a diameter of 1 in.

The radiation resistance can be further increased by mutual coupling, for instance, (1) when multiple N turns are used in a loop, (2) when the loop is wound on a magnetic material, and (3) when a design employing multiple N inductively coupled radiating elements is used. The latter case pertains to the Meander line and the Goubau antennas, where $N = 2$ and 4, respectively. The radiation resistance is raised by N^2, allowing the resistive portions of the antenna to be larger, resulting in smaller conductor area for a given efficiency.

The reactive portion of the antenna impedance seen at the terminals is capacitive for a dipole and inductive for a loop. The dipole requires a series inductor to cancel the reactive portion of the antenna impedance, that is, for resonance/impedance matching. The loop typically requires a series capacitance to cancel the inductive reactance at the frequency of operation, that is, to create a resonance condition where the radiation resistance plus loss resistance is seen at the input terminals. For the dipole, the inductor or so-called "loading inductor" is best placed above the base of the antenna, as the current is high at the base where I-squared R losses in the coil are more substantial.

For high radiation efficiency, the radiation resistance should be much larger than the loss resistance. For broad bandwidth, the radiation resistance should be large in relation to the reactive part of the unmatched antenna impedance (the stored energy) as required by the definition of Q.

For the most part, the monopole, dipole, and loop antennas thus far discussed have been replaced by much shorter embedded antennas in today's wireless devices. Many of the basic principles governing the antenna efficiency and bandwidth can still apply but are modified by the effects of the ground plane which acts as a *counterpoise* and often the primary radiator with the antenna element acting as a resonant coupler exciting characteristic modes thereupon for the driven element.

4.3 ANTENNA REQUIREMENTS: CURRENT AND FUTURE—BAND COVERAGE, RADIATION EFFICIENCY, COUPLING, CORRELATION, AND ARCHITECTURAL VARIANTS

The factors driving the next wave of wireless device designs are fairly straightforward—more functionality and more power in a smaller package at a lower cost. Accomplishing this goal is more difficult than it appears. As designers try to meet these requirements they have had to take a close, critical look at every component within their designs in an attempt to squeeze more desirable characteristics out of each element—batteries need to last longer, electronic components need to do more functions while consuming less power, and all device elements need to be more compact in size. But one element that has received less attention throughout the wireless revolution has been the very element that makes a device wireless—the antenna. Until recently, antennas have been incorporated into wireless devices with little modification to the traditional monopole and loop designs that have been used for decades. Today, however, as companies compete for ever-smaller devices with increased functionality, the search for new, more efficient antenna configurations has once again become a factor in wireless designs.

4.3.1 Factors Driving Antenna Design

There are many factors driving innovations in antenna design. The initial factors, such as smaller size or added frequency bands, are consumer driven by the market. However, these market factors create other technical requirements such as maintaining or improving performance with less battery consumption. All considered, these competing factors present wireless designers with a perplexing challenge to resolve. A quick survey of the most pressing requirements driving antenna design will show the extent of this wireless "Gorgian knot."

4.3.1.1 Reduction in Size

Perhaps the most obvious antenna characteristic to consumers, antenna size has shrunk along with other handset components as a whole. Antennas on mobile phones have transitioned from whips, to retractable whips, to stubbies, and now to fully embedded antennas. This reduction in obtrusiveness has been driven by the overall quest to reduce the size of wireless devices, especially handsets.

But with the introduction of the antenna to the inside of the phone casing, new problems arise. The most obvious is that as the antenna moves from an external to an internal component, designers now have to devote valuable internal real estate to a new element. And this is occurring just as designers are trying to reduce the size of everything inside the casing. To complicate matters, immutable laws of physics require certain components, such as the antenna and the battery, to maintain a minimum size to meet bandwidth and energy requirements.

Also, the performance of the handset may drop dramatically depending on the hand position or other environmental factors. This is due mostly to the absorption of the antenna signal by the user's body, especially the head and hand. While a quick fix is to attach an extended earpiece/microphone, this is not always convenient for the consumer since the additional wire is cumbersome and gets in the way of many activities. Additionally, the specific absorption rate or SAR regulatory limit sets the maximum energy absorption allowable to the body.

4.3.1.2 Increased Radiating Power

In order to combat the loss of signal strength due to embedding the antenna inside the device casing, many handset manufacturers are calling for the smaller antennas to exhibit greater radiating performance. Antennas frequently exhibit a loss of up to 2 or 3 dB once they are inserted into the casing. However, customers will not keep a phone that continually drops calls, no matter how stylish it looks. So designers are tasked with finding a way to overcome the detuning effects of the casing *and* of the user's body in order to maintain acceptable signal strength. Adaptive tuning of the antenna is one way that is being developed to mitigate these effects.

4.3.1.3 Decreased Battery Consumption

While increasing radiating power is possible, device manufacturers do not want to use more battery power to do it. The transmitted signal can consume more battery power than any other operation by a handset device. So designers are facing a contradictory dilemma—radiating more power while using less power. A factor in antenna design that already deals with this relationship is the antenna's *efficiency*. *Radiation efficiency* is a measurement (as a percentage) of an antenna's ability to convert applied electrical energy into a radiated electromagnetic energy. The more applied energy the antenna transforms into a radiated signal, the more efficient it is. However, embedded antennas typically exhibit efficiency of about 50% or less especially when considering absorptive and detuning due loading by human tissue. While some of this poor performance can be attributed to the handset casing and other lossy components near the antenna mentioned above, designers must strive to develop antennas and radio frequency (RF) systems that are as practically efficient as possible in order to get the most out of both the device's battery power and available RF power. Tuning the antenna matching network or centering the antenna resonance to the actual operating frequency through active control or tuning offers the potential to do just that.

4.3.1.4 Higher Gain

Gain is a combined measurement of the antenna's *efficiency* and *directivity*—the parameter describing the directionality of the emitted RF signal radiated in a particular, focused direction. Many people assume that the average handset antenna emits its signal equally in all directions, giving the antenna a perfectly spherical beam pattern. However, this is not the case, as the embedded antenna in reality acts as an exciter for radiation currents that exist on the wireless device circuit board or "ground" structure. Even antennas that are designed to have spherical patterns exhibit shapes that may resemble a doughnut due to the interaction with the circuit ground. Additionally, antennas can emit complex lobed patterns, giving the signal pattern an irregular shape. What all this means to the designer is that an antenna that is very efficient at sending a strong signal in one or more particular directions can help overcome the power emission/consumption issues stated above, and in particular, at some frequencies some degree of beam pattern control can be achieved through adaptive control of the antenna with parasitic elements or by phasing of multiple feedpoints on the antenna itself.

4.3.2 THE EVALUATION OF ANTENNA PERFORMANCE: DESIGN ISSUES

These new requirements being demanded of mobile antennas have made antenna Engineers must take a closer look at how antennas have been traditionally evaluated for commercial use. The traditional parameters of gain, voltage standing wave ratio

(*VSWR*), frequency, and bandwidth are no longer enough to determine an antenna's effectiveness for use in a new product design and designers now must consider the factors of antenna efficiency and volumetric size in order to achieve optimal RF performance with their new wireless products.

4.3.2.1 Traditional Requirements

4.3.2.1.1 VSWR

Technically, the *VSWR* is a measure of the *return loss* of an antenna, which is the difference between the power input to and the power reflected from a discontinuity in a transmission circuit. This loss occurs due to an impedance mismatch between the transmission line and the antenna. This power is retained inside the circuit and not available for broadcast and is clearly a factor contributing to the inefficiency of the antenna system. In a perfectly matched transmission system, there are no standing waves and the *VSWR* (a ratio metric measure of the crest to null of the voltage standing on the line) is 1.

Generally, antennas having *VSWR* less than 5:1 are acceptable for receive applications and low power transmission, as some loss is tolerable. High power transmission, however, often requires a *VSWR* less than 3.5:1 to avoid damaging the power output circuit, and is ideally less than 2:1 in modern wireless devices.

RF designers sometimes prefer measuring *return loss*, which is 10 times the logarithm of the ratio of the incident RF power to the reflected power, that is

$$RL = 10\mathrm{Log}\left(\frac{P_{incident}}{P_{refelected}}\right) \tag{4.6}$$

Return loss is related to *VSWR* according to

$$RL = -20\mathrm{Log}\left[\frac{(VSWR-1)}{(VSWR+1)}\right] \tag{4.7}$$

For handset antennas in the past, *RL*s of −5 dB (*VSWR* ~ 3.5:1) has been acceptable, but with the advent of cost effective tunable components for use in the matching network much better *RL* can be feasibly obtained dynamically at the operating frequency.

For example, transmitted power at *VSWR* 5:1 is about 55%, while at 2:1 is 89%.

4.3.2.1.2 Frequency Coverage and Bandwidth

Frequency coverage and antenna bandwidth, the range of frequencies within which the antenna performs to a determined standard, continue to expand. The frequency range over which most antennas operate at fundamental resonance is usually rather narrow—about 10% of the center frequency is fairly common, and this depends on the *Q* of the entire radiating system.

It is common for antennas to be designed that have multiple resonances—each for a corresponding frequency band. In the latter case, the antenna may be constructed with multiple distinctive elements: one for each band at a maximum. Antenna designers are then faced with the complex task of correctly forming the size and spacing of the elements and matching their resonant impedance (voltage-to-current ratio). One of the biggest problems encountered in this process is that each element acts as a parasitic load on each of the others and, therefore, the tuning of one element has a noticeable effect on the others. Once the elements are properly constructed, the antenna sends and receives RF signals on different frequencies each corresponding to the resonance of each of the elements. Assuming that the antenna does not need to operate on all bands simultaneously, a scheme using adaptive tuning of the front-end electronics to the antenna impedance may be employed to mitigate the antenna design problem of impedance matching over all bands simultaneously.

4.3.2.1.3   Bandwidth

Bandwidth is the amount of electromagnetic spectrum needed or allocated for a particular communications channel or group of channels. It is usually defined in units of frequency and is computed as the difference between an upper and lower band edge limit. For instance, Band 2 (Table 4.1) is defined as 1850–1990 MHz, and, therefore, has an allocated bandwidth of 140 MHz. Individual channel bandwidths, however, are much narrower requiring only 10's of kilohertz bandwidth.

Most antennas used in handheld devices are *multiband*—they only operate on the band of frequencies for which the device was intended. *Broadband* antennas tend to perform less effectively than narrowband antennas and typically cost more to produce. However, the advantages offered by broadband antennas (covering multiple frequency bands, ability to transmit more data faster, etc.) can make them attractive for wireless device makers. The key is finding a happy medium between performance and bandwidth coverage, and in some cases depending on the front-end electronics architecture, multiple feedpoints are possible thereby splitting operation between the low bands and the higher bands. Such architectures using more than one antenna can be effective in reducing losses in both the front end and the antenna, but at the expense of extra feed lines and potential assembly cost. Increased antenna space or volume can also be a factor.

The increase in the demands now expected of antennas greatly increases the importance of two characteristics—efficiency and size.

4.3.2.1.4   Efficiency and TRP

The *radiation efficiency* parameter has become an accepted means for comparing antenna performance in many handheld applications, as it is a direct measure of how well an antenna transforms onboard electrical energy into transmitted signal energy. This is a concern in achieving high performance while preserving precious onboard battery energy. A 100% *efficient* antenna would theoretically convert all input power into radiated power, with no loss to resistive or dielectric elements.

The demand for greater signal strength while consuming less battery power makes an antenna's efficiency all that much more important. This is especially true in the case of embedded antennas that lose some signal strength in the casing and nearby components.

While antenna designers have traditionally used antenna efficiency as a design requirement, most cellular operators specify the minimum *total radiated power* or *TRP*, which is typically measured in dB over 1 mW (dBm). Therefore, it is the combined system of antenna efficiency, front-end filters, switch, feed line loss as well as power amplifier output that must meet minimum specification on a band-by-band basis. Therefore, the concept of adaptively matching these elements by tuning for optimum performance can make sense from a power efficiency standpoint.

4.3.2.1.5 Noise Coupling, TIS, and Antenna Location

In addition to antenna efficiency that directly affects necessary transmit power amplifier output to achieve a target *TRP* value, antenna placement relative to conducted noise currents or radiated noise sources also matters for adequate receiver performance. Typically, wireless operators specify the noise floor in terms of a minimum parameter called the *total integrated (or isotropic) sensitivity* or *TIS*.[7] It is a measure of the average measured sensitivity of a receiver-antenna system averaged over a 3D sphere. The result will be strongly related to the antenna efficiency, the radio noise floor, and the presence of noise sources that are coupled to the antenna circuit.

The system under test (the receiver plus antenna) is operated in active mode using a *Base Station Emulator* that transmits to the system under test with an auxiliary antenna. The emitted power is lowered until the *Bit Error Rate* or *BER* reaches a power threshold at the receiver (measured in dBm). This is done for a fixed azimuth and elevation angle to arrive at an *effective isotropic sensitivity* (*EIS*) for that specific angle and wave polarization. The total isotropic sensitivity (*TIS*) is then the given by *EIS* components averaged over the entire sphere:

$$TIS = \frac{4\pi}{\int_0^{2\pi} \int_0^{\pi} [(1/(EIS_\theta(\theta_1,\phi_1))) + (1/(EIS_\phi(\theta_1,\phi_1)))] \sin\theta_1 d\theta_1 d\phi_1} \tag{4.8}$$

Unfortunately, there is often limited opportunity to relocate antennas or noise sources in a production design process. Control of antenna frequency as well as bandwidth using adaptive tuning in conjunction with adaptive frequency control of the interferer can provide better rejection and a means to more easily meet *TIS* requirements.

4.3.2.1.6 Future Requirements

Handsets and other wireless devices have in the past relied on either single antennas with one feed connection or dual antennas with one primary antenna and one secondary antenna operating as a diversity, or receive only antenna. As such, the secondary antenna could operate with reduced bandwidth and efficiency reduction of 3 dB or more compared to the primary antenna. With recent emphasis on 4G and LTE Advanced,[8] supporting MIMO operation in the wireless device or *user equipment* (UE) is becoming necessary. Higher order MIMO 2×2 or greater is seen as viable relief for unloading already crowded networks. To achieve full potential, the

UE must support at least downlink capability with two receiving antennas of similar gain and efficiency. Future plans as indicated by 3GPP and some wireless service providers suggest that as many as four antennas will be required in wireless handsets and tablet devices. This imposes new requirements for antenna efficiency, coupling, and signal correlation between antennas.

4.3.2.1.6.1 Antenna Efficiency and MIMO   In the case of handsets uplink MIMO with two or more antennas need to meet antenna efficiency requirements similar to those of currently produced devices. Radiation efficiencies of 50% or greater corresponding to an average gain of −3 dB or greater are common, and likely to be more demanding at higher frequencies such as Band 7 where greater propagation loss is typical. This can be a significant challenge as placement of antennas affects the antenna efficiency due to feed line loss and antenna-to-antenna coupling.

4.3.2.1.6.2 Antenna Coupling   For many MIMO systems, antenna coupling greater than −10 dB is considered objectionable. Several factors drive this specification and include (1) reduction of the signal independence as determined by one or more measures of *correlation coefficient*, (2) power loss into opposite port load terminations affecting power amplifier performance and linearity, (3) component power dissipation overloading, and (4) efficiency loss and desensitization of receive circuits.

4.3.2.1.6.3 Antenna Correlation   Several measures for signal correlation between antenna ports are common. One of the most common is the *antenna pattern envelope correlation coefficient* or *ECC*. It is typically measured using antenna pattern measurements, which classically are done by measuring the spherical field components radiated from one antenna while applying normal loads to the opposite antenna(s). The roles of two antennas are then switched to obtain another set of field data without moving either antenna. From the data sets collected, the pattern correlation coefficient can be calculated from full sphere antenna pattern measurements and is given in terms of the electrical field components by[9]

$$\rho_p = \frac{\displaystyle\int_0^{2\pi}\int_0^{\pi} A_{12}d\Omega}{\sqrt{\displaystyle\int_0^{2\pi}\int_0^{\pi} A_{11}d\Omega}\sqrt{\displaystyle\int_0^{2\pi}\int_0^{\pi} A_{22}d\Omega}}$$

where $A_{mn}(\theta,\phi) = XE_{mv}(\theta,\phi)E_{nv}^*(\theta,\phi) + E_{mh}(\theta,\phi)E_{nh}^*(\theta,\phi)$ (4.9)

and the cross-polarization power ratio $X = S_v/S_h$

The subscript indices (m,n) refer to either antenna port 1 or 2, depending on their integer value, $\Omega(\theta,\phi)$ is the spatial angle in steradian as commonly depicted, and $E_{1,v}$, $E_{1,h}$, $E_{2,v}$, and $E_{2,h}$ are the complex envelopes of either the vertical or horizontal field components resulting from excitation of either antenna port 1 or 2, respectively. The cross polarization ratio is defined as the ratio of the mean received power in the vertical polarization to the mean received power in the horizontal polarization.

Since the signals may arrive from different directions, the antennas should exhibit characteristics that allow for highest independence of the received signals. A probability function P may also be used to weight the computation to match the angular arrival distribution expected, for the expected scattering functions associated with the propagation environment. The method involves the use of a simple environmental model that has been experimentally verified.[10] In this model, the signal power is described as a function of angle of arrival by probability functions P_v and P_h, which can be included in Equation 4.8 as follows:

$$A_{mn}(\theta,\phi) = XE_{mv}(\theta,\phi)E^*_{nv}(\theta,\phi) + P_v(\theta,\phi) + E_{mh}(\theta,\phi)E^*_{nh}(\theta,\phi)P_h(\theta,\phi) \quad (4.10)$$

Difficulty achieving low values of ECC ($\rho^2 < 0.5$) depends on factors such as antenna separation, polarization, and currents induced on the ground structure. In the latter case, the structure can be analyzed to determine its modal behavior, and how each antenna placement excites or contributes to existing in a particular mode.[11]

4.3.2.1.6.4 Antenna Isolation and Correlation Antennas should not only exhibit low port-to-port signal correlation, but should also exhibit isolation from one another so that signals being broadcast or received from one feed port don't appear at the opposite port or feed. The isolation is typically measured at the antenna terminals by using the commonly derived S-parameters. Isolation figures are typically given in decibels and are designated as S_{12} or S_{21} depending on whether the measurement is from antenna 1 to 2 or vice versa. If the isolation is poor, the source antenna will deliver substantial power into the adjacent antenna's termination impedance (typically 50 ohms-resistive), with a corresponding reduction of the overall radiation efficiency. S_{21} values of -10 dB are regularly deemed acceptable, resulting in power loss of less than 10%.

Since the antenna near-field coupling is related to the far-field pattern, it is possible to approximately express the correlation coefficient in terms of the S-parameters measured at each antenna terminal. This is given by[12]

$$|\rho_s| = \frac{|S^*_{11}S_{12} + S^*_{21}S_{22}|}{(1-(|S_{11}|^2 + |S_{21}|^2)^{1/2}(1-(|S_{22}|^2 + |S_{12}|^2)^{1/2}} \quad (4.11)$$

Comparison of measurements taken by this approach can approximate those computed using the pattern method, and are much simpler. From this equation, however, it is clear that small values for S_{12} and S_{21} can result in small values for ρ_s. There are, however, cases where the correlation coefficient might be small even if S_{12} and S_{21} are larger, depending on the input match reflected in the S_{11} and S_{22} values.

Measurement data have been taken to define parameters for Gaussian forms of P_v and P_h in various environments.[13] This data supports the general methodology for computing ρ based on a field model (versus that of pattern alone) with good correlation between the two computational approaches. Perhaps most interesting is that the

Tuned Antennas for Embedded Applications

correlation coefficient depends primarily on the phase difference between the two antenna patterns—indicating that good correlation performance requires careful consideration of the antenna placement and design in the case where multiple single antennas are used in close proximity.

Briefly then, the antenna system, its associated pattern, and the environment are all clearly important in establishing independence between the signals received at each antenna and, therefore, the overall system data transfer rate.

4.4 BASIC ANTENNA TYPES: IFA, PIFA, MONOPOLE, RING, AND SLOT

Although many types of wireless devices depend on antenna technology, the designer generally uses a toolbox comprised of several basic antenna types. Depending on the space and location allotted in the industrial design, a specific antenna type might be chosen for initial trial either by simulation or physical measurement. The basic antenna types and their variants comprise the majority of designs.

4.4.1 MONOPOLE

Monopole antennas rely on a single radiator and a corresponding *counterpoise* as shown in Figure 4.2a. In the case of a typical wireless handset or other wireless device, the ground structure and printed circuit board (PCB) ground plane establish the counterpoise, while the radiator may be a printed conductor on a variety of substrates, including the case. The radiator may be "loaded" with lumped or continuous elements that exhibit an electrical reactance that cancels an opposite equivalent reactance associated with the combined structure at the frequency of interest. The monopole exhibits a theoretical resistance that is one half that of a dipole (see Equation 4.4) at the same design frequency and can be perpendicular or parallel to the PCB ground plane—the latter is called a *transmission line monopole*.

4.4.2 IFA/PIFA

The *inverted F* or *IFA* antenna is shown in Figure 4.2b and is a variant of the *transmission line monopole* and includes a termination to the counterpoise structure. The position of this termination relative to the feedpoint allows matching of the antenna impedance at resonance to the feedline.

When the antenna width is increased (Figure 4.2c) to form a planar structure, the antenna is termed a *planar inverted F* or *PIFA*.

Either antenna type shows high concentration of current in the feed to ground point and, therefore, high near field RF intensity.

4.4.3 RING

The ring antenna is a recently used antenna type that surrounds the ground plane associated with the wireless device. It has become popular as it interferes minimally

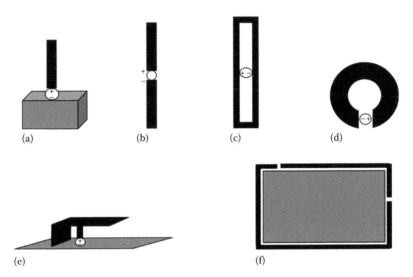

FIGURE 4.2   Examples of typical antenna types: (a) Monopole, (b) dipole, (c) slot, (d) loop, (e) PIFA, and (e) ring.

with components on the PCB and can be designed somewhat independently. It is usually driven at some point on the ring structure and grounded at one or more points along the ring. Functionally, it excites resonant modes on the ground plane but can be subject to hand or other loading at high impedance points along the structure.

4.4.4   SLOT

The slot antennas can be considered an analog of dipole antennas, having magnetic and electric fields interchanged. Typically, the slot is formed between two conductive planes and is a finite length that determines the fundamental resonant frequency. An electric field is established from one side of the slot to the opposite side and shows decreasing magnitude from the feed to the terminus. Slot antennas are used in predominantly metal structures and may be fed with a probe near the centerline.

4.4.5   LOOP ANTENNAS

Loop antennas that take a variety of shapes are fed at a break in the conductive loop structure. Another property of the loop is the ability to achieve multimode operation in one antenna structure. Square loops can be used in two radiation modes that can be excited simultaneously or that can be individually selected electrically. The fundamental mode is that of a monopole producing a linear polarized radiation field resembling that of two closely spaced monopole antennas. In the monopole mode, the primary antenna current is perpendicular to the antenna ground plane. A second mode of operation is the "loop mode" where the primary antenna current is parallel to the antenna ground plane and the maximum radiation is perpendicular to the ground plane.

4.4.6 ADAPTIVE ANTENNAS

Also commonly known as "smart" antennas, these are antennas that focus their RF energy from the mobile handset to the base station or that direct a radiation pattern null toward an interferer.

Adaptive antenna arrays have two great advantages: (1) they allow more spectrum to be reused because omnidirectional signals are not overlapping and (2) they can send a strong directed signal using less energy because energy is not wastefully radiated in directions it is not needed. In addition, the mobile device will also require less energy and conserve battery power in some systems.

4.5 TUNING METHODOLOGY AND SYSTEM CONCEPTS: APERTURE-BASED, MATCH-BASED, OPEN LOOP, CLOSED LOOP

Various tuning methods are possible and can be classified according to function. *Aperture-based tuning* provides direct control of the antenna's resonant frequency through the change of the effective length of the antenna and can be accomplished through either physical or electrical means. In the latter case, either the inductive or capacitive reactance can be altered via different architectural means using tunable capacitors or inductors. *Match-based tuning* is intended to be functionally part of a matching network and can be located anywhere in the signal RF chain where a mismatch in impedance is likely to occur, for instance, between PA and filter, bank/switch or at the antenna feed terminal. In the latter case, because the antenna can be modeled as a lumped element network, both the resonant frequency of the antenna and the match (return loss) can be subject to control.

Various antenna models have been developed to accurately model the antenna resonant frequency and input impedance as a function of frequency,[14] and several will be covered here.

The simplest model is a series RLC circuit as shown in Figure 4.4a for a dipole at fundamental resonance, where

$$Z_{in} = \frac{1}{sC} + sL + R$$

As can be seen in Figure 4.3b the actual behavior of an antenna is only approximated in this example as from Equation 4.4, the resistive component should increase with frequency as shown. Nevertheless, the circuit is illustrative of the need for adaptive tuning if operation is desired at even 10% above or below resonance (shown here as a normalized frequency of 5). Also note that an inserted series capacitance or switched circuit inductance would change the resonance frequency but that the real part of

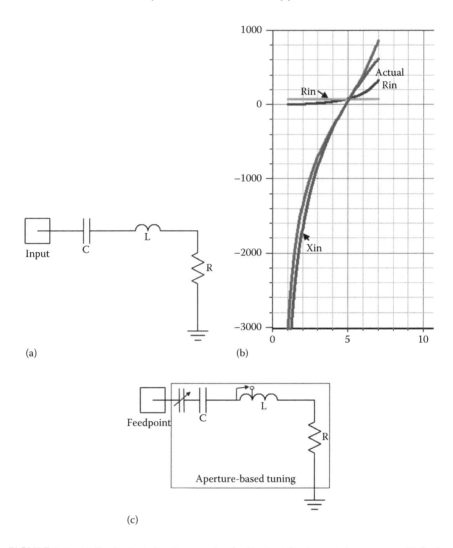

(a)

(b)

(c)

FIGURE 4.3 (a) Equivalent circuit example of a dipole at fundamental resonance. (b) Series circuit model versus actual real (Rin) and imaginary (Xin) components of antenna impedance. (c) Possible tuning methodology, for example, of Figure 4.5a.

the input impedance would still be mismatched as also can be seen from Equation 4.4 and Figure 4.3b. The need for *match-based tuning* compensating for the change in the resistive component of Z_{in} might also be called for in some applications. So from this simple example, it can be seen that a series variable capacitance within the antenna could result in raising the resonant frequency and that a series selectable inductance would have the opposite effect (Figure 4.3c). Note also that either component could be considered as part of a matching network external to the antenna (outside the black outline).

The previous example is illustrative, but real world antennas are complex multi-resonant devices. As an example, the circuit model shown in Figure 4.4a is exemplary of a broadband multi-resonant antenna developed by Foster.[15] The broadband model is reasonably accurate over the three resonant frequencies shown.

Such models can be used to develop a tuning methodology based on basic circuit design principles. The model parameters can be matched to a particular antenna design and/or a new model may need to be developed that includes a lumped model tuning elements if available. Once a candidate tuning architecture is established, a physical model of the antenna and tuning components can be simulated using one of the many 3D EM simulators available commercially to refine the design.

4.5.1 Open-Loop Tuning

There are various motivations throughout the industry for using tuned circuits in the front-end architecture of a wireless device. Expectations among wireless service providers are that device RF performance would improve as tuning theoretically has the ability to mitigate impairments due to absorptive hand loading and other environmental usage conditions. Component suppliers for power amplifiers, antennas, and other components see a potential for reducing cost or design constraints that result in loss of power efficiency. Original equipment manufacturers or OEMs see a potential for reduced space usage as tuned antennas can have higher Q. This is particularly true in the low-frequency region where the phone ground structure is the primary radiator and the antenna is a transducer that couples RF energy to the structure. Each of these separate motivations can drive the tuning architecture toward different realizations.

Figure 4.5 illustrates the case where simple open-loop control of the antenna resonance is desirable to cover four low-frequency bands but with smaller volumetric sizes than with a broadband antenna alone. Open-loop control requires only information from the baseband processor concerning the frequency of operation. To illustrate that a volumetric size reduction is possible, two different antenna configurations were analyzed using a 3D EM modeling program, one being a broadband design and another a narrower band design capable of tuning over the same frequency range but using four tuning states. Figure 4.5a illustrates a $50 \times 6 \times 14$ mm seven-band antenna configuration and its associated radiation efficiency over just the lower three-band spectral region from 700–960 MHz. A similar, but smaller ($50 \times 6 \times 7$ mm) antenna configuration is shown in Figure 4.5b illustrating, that tuning using only four states is able to produce nearly the same efficiency, and total frequency coverage as the larger broadband antenna.

In this example, it is clear that a physical volume reduction of ½ can be achieved by tuning the antenna to one of several states, each supporting a certain set of frequency bands. The antenna during operation would, therefore, only be required to change state when the operating band is changed. The time required for this change must be compatible with other functions within the radio system. A typical requirement might be 10–20 microseconds or less.

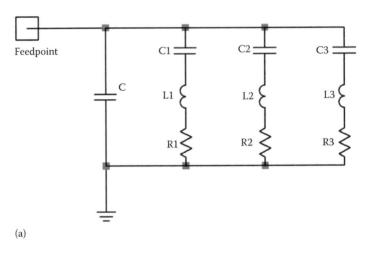

(a)

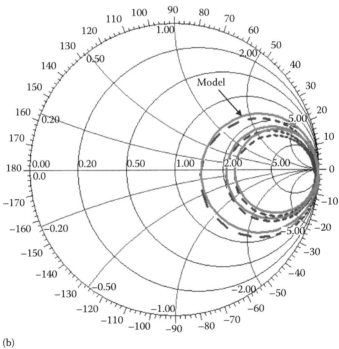

(b)

FIGURE 4.4   (a) Foster's Second Canonical Model with small added losses. (b) Smith chart plot of actual dipole impedance versus model of Figure 4.3a.

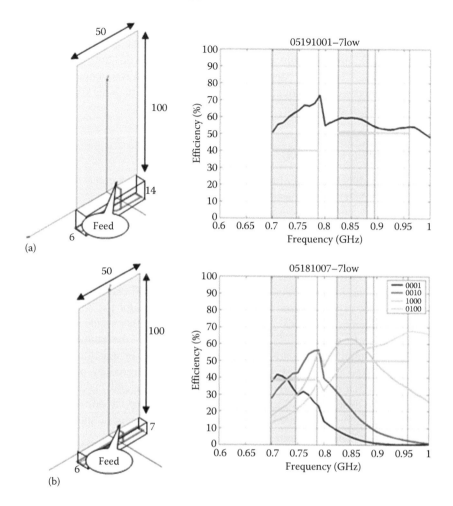

FIGURE 4.5 Comparison of (a) multiband antenna and (b) tuned antenna with respect to size and radiation efficiency over the region 700–960 MHz. (Dimensions in mm.)

This is just one example, and the antenna must still be able to operate over a much broader bandwidth characteristic of the many air interfaces supported by the mobile device—typically 700–2600 MHz in a 4G-world phone. It is, therefore, necessary to tune the antenna so that its instantaneous bandwidth at any time is sufficient to cover the expected operating frequencies associated with the usage model. This is particularly true for LTE-Advanced where *interband carrier aggregation* (CA) requirements are such that the antenna and radio system must operate simultaneously on multiple bands at the same time. For example, a phone operating at B17 (state 0001 in Figure 4.8) may also need to operate in B4 (see Table 4.1) for CA 17–4 according to 3GPP requirement. Although this requirement could potentially negate the use of tuning to separate bands, it is important to note that many CA band combinations utilize low-band–high-band operation allowing the antenna designer to separate the low- and high-band elements of the antenna. Since achieving a broad bandwidth low Q antenna is particularly difficult for the low bands (600–960 MHz), tuning might be applied to only the low-band element while achieving broader fractional bandwidth at the higher band frequencies. These are details associated with a particular implementation, and are generally beyond the scope and intent of this chapter.

4.5.2 CLOSED-LOOP TUNING

Closed-loop feedback control of the antenna frequency or impedance matching network is used when conditions that may cause detuning or adverse operation are unknown to the onboard processors. Such control schemes can be either local or remote; that is, under direct control of the local device or commanded from received information obtained remotely and sent back to the local device. Motivation for closed-loop control is primarily to improve system performance and can take many forms.[16] Local control may use onboard sensors to determine device orientation and/or hand location from which a tuning command signal can be generated from a lookup table or expert system. It is also possible to monitor the *return loss* from the transmit antenna and make adjustments to minimize that signal by tuning the matching network. Alternatively, it is possible to "sniff" the emitted RF signal and make adjustments to the antenna resonant frequency or matching network to maximize the near field signal. Other schemes, some covered by patents, derive frequency information or bandwidth information directly and make adjustments accordingly. An example of a local closed-loop system used for optimally tuning a 2×2 MIMO antenna is illustrated in Figure 4.6. Coarse frequency control is afforded

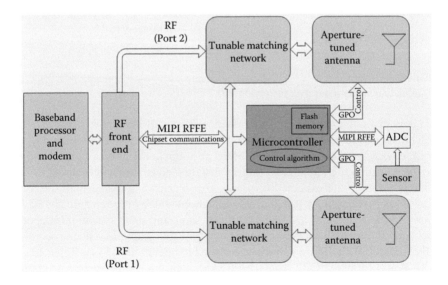

FIGURE 4.6 Block diagram local closed-loop control of a demonstrator MIMO antenna.

using a *tuned-aperture* method for band-to-band operation while fine tuning is accomplished with a *tuned-match* approach. Many different algorithms are possible depending on sensed parameters and the types of control used and are beyond the scope of the section.

4.6 TUNING COMPONENT REQUIREMENTS: LOSS MECHANISMS, PARASITICS, AND NONLINEARITY

Tuning components can exhibit a variety of non-ideal behaviors that oppose their potential for improving the antenna or device performance. In general, the issues are complex and depend on the tuning methodology used, the device specifications, physical structure, and composition, as well as packaging.

4.6.1 LOSS MECHANISMS

There are a variety of loss mechanisms that contribute to both antenna and tuned component losses.

> *Resistive near field coupling* is due to radiated fields from the antenna circuit inducing current into nearby components that exhibit resistive dissipation. Components exhibiting such loss mechanisms may be either passive or active devices and may, therefore, be either linearly or nonlinearly related to the field amplitude.
> *Conducted resistive loss* is attributable to I^2R losses in the circuit carrying the antenna radio frequency currents. These losses are related to the antenna conductor geometry, composition, and frequency as well as any series or parallel resistive losses resulting from tuning components that are in the front end or tuning circuits. Therefore, the Q of the tuning components can become a sizable factor in determining the overall benefit or gain of tuned versus passive antennas.
> *Nonlinearity loss* stems from components that exhibit a higher order dependence on either the voltage or current at their terminals. This then converts the desired excitation in one spectral region to undesirable frequency content. As such, it represents some power loss in the desired spectrum but more importantly can mix down band into adjacent channels. Typical measures used are second and third harmonic relative to the carrier frequency as well as IP2 or IP3 measures.[17]

4.6.2 PARASITICS

Parasitics are couplings or parameters that depart from idealized models. *Parasitic capacitance* for instance can be particularly troublesome in tuning circuits and may result from stray capacitance from the semiconductor die to power or ground connections, or from interelectrode capacitance. This often results in a reduction of the range of tuning such as the C_{max} to C_{min} ratio. Additional parasitic coupling can result from RF coupling from the antenna conductors to the tuning component die itself. Depending on the induced field amplitude, the tuning device may exhibit nonlinear capacitance and may not allow normal performance such as the selection capacitance values at the active terminals. For this reason, tuning circuits must be tested at full applied power, and may require individual shielding. Also, package selection is critical for some applications.

From the antenna perspective, parasitic capacitance from the tuning terminals to the control circuits within the tuning chip can be particularly problematic at the highest antenna frequencies where even small capacitances on the order of 1 pF or less exhibit low enough reactance to increase loss to intolerable levels. For instance at 3.5 GHz, 0.5 pF exhibits a reactance of 90 ohms providing a substantial coupling to other circuit elements when used in a 50 ohm circuit. Designers can avoid these effects by careful design of the antenna and tuning architecture.

4.7   TUNED ANTENNA EXAMPLES

Many examples exist for tuned antennas and encompass both *match- and aperture-based* methods. These have evolved beyond the R&D stage and are in production in a number of handsets currently (in 2015). Details in some cases are closely guarded until production release, and various devices are available for use as noted in other chapters of this book.

For the purpose of illustration, two examples will be covered showing single feedpoint antennas that cover the 700–2700 MHz spectral range.

4.7.1   EXAMPLE 1: APERTURE-TUNED HANDSET ANTENNA

Illustrated in Figure 4.7 are two prototype antennas that meet specifications as provided by a major international wireless operator for diversity or 2 × 2 MIMO operation.[18] Each antenna consists of a small PCB (35 × 6.5 × 0.8 mm) that includes all tuning necessary to cover the bands B17, B13, B5, and B8, as well as the high bands above 1 GHz (1710–2170, 2300–2700 MHz). The 6.5 mm height is driven primarily by the tuning chip package and could be reduced using the latest *chip scale packages* or CSPs. The antennas are positioned at each end of a 50 × 120 mm PCB with form factor similar to many handsets.

The circuitry shown on the main body of the PCB consists of battery and decoders for driving the tuning chip onboard the antenna, as well as transmission lines and SMA connectors to allow RF performance testing of the system. Performance goals for the antenna system are shown in Table 4.2.

Each antenna is separately controllable and configured to use three-tuning states, which are necessary to achieve performance in each of the lower bands. The average efficiency per band is shown in Figure 4.8 for each of the antennas. Note that the primary and secondary antennas are specified with different gain requirements each represented with different shading (left most bar per pair: primary antenna at port 1).

FIGURE 4.7 Aperture-tuned prototype antenna system.

TABLE 4.2
Performance Goals for a Dual Antenna System

Parameter	Requirement	Units	Comments
# Antennas	2		Primary, Secondary
Frequency coverage	B17, 13, 5, 8, 3, 2, 39, 1, 34, 40, 7, 38	Band	
Gain, Primary, High bands	>−3 primary >−6 secondary	dBi (average)	
Gain, Primary, Secondary	Per curve	dBi	
Isolation	<−10	dB	
Power (max)	> +25	dBm	
VSWR	<3.5:1		All bands
Correlation coefficient	<0.7 <0.5 high bands		
Size (ground plane)	50 × 115	millimeters	
Antenna volume	<1600	mm³	Including ground keep-out
Control voltage	2.7	Volts	Nominal
Control interface	Parallel (2-bit) or serial		

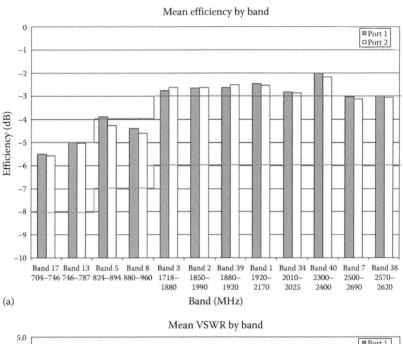

(a)

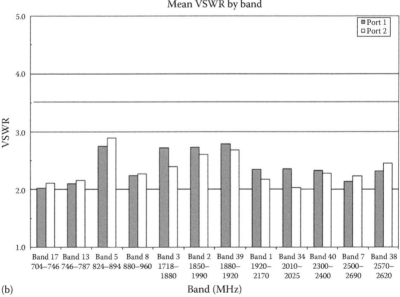

(b)

FIGURE 4.8 (a) Measured antenna performance for primary and secondary antennas shown in Figure 4.9. (b) Mean values of *VSWR* for each band. *(Continued)*

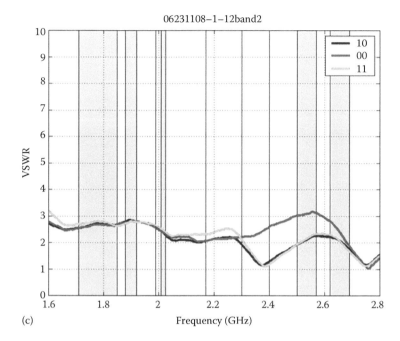

(c)

FIGURE 4.8 (Continued) (c) Measured high-band *VSWR* versus tuning state (10, 00, 11) showing minimal impact of low-band tuning on high band.

As can be seen in Figure 4.8c, the low-band tuning has a minimal effect on the high-band *VSWR* so that various CA band combinations would be possible (the tuning state is shown as a binary number 10, 00, or 11).

4.7.2 EXAMPLE 2: MATCH-TUNED HANDSET ANTENNA

In this second example, the band coverage and RF performance requirements are identical to the previous case. The form factor, however, is for a USB dongle with dimensions 23 × 35 mm which would be used in conjunction with a large ground structure typical of a notebook PC. The design incorporates a dual feed single element antenna using iMAT™ technology[18] tuned via two *digitally tuned capacitors* or DTCs that can be of MEMs or semiconductor type. MEM was chosen for initial implementation in the matching network due to its availability at the time the prototype was constructed. The prototype hardware shown in Figure 4.9 uses a microcontroller to generate serial peripheral interface (SPI) bus signals required by the DTC. The tuning state is selected sequentially using pushbuttons on the PCB.[19]

The advantage of using DTCs in this application is the large number of selectable states (32–64) controllable via the SPI or RFFE bus. This allows fine-tuning of the antenna match and frequency in response to production tolerances or due to environmental loading. Shown in Figure 4.10 are efficiency plots for five open-loop tuning states used. Here, the DTC separately controls two capacitors in each port's matching network.

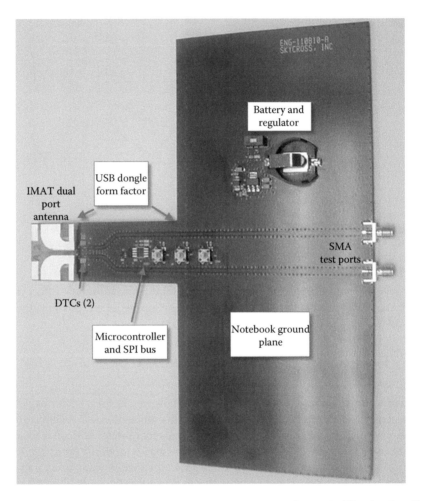

FIGURE 4.9   USB dongle dual port aperture-tuned antenna using IMAT™ technology.[18]

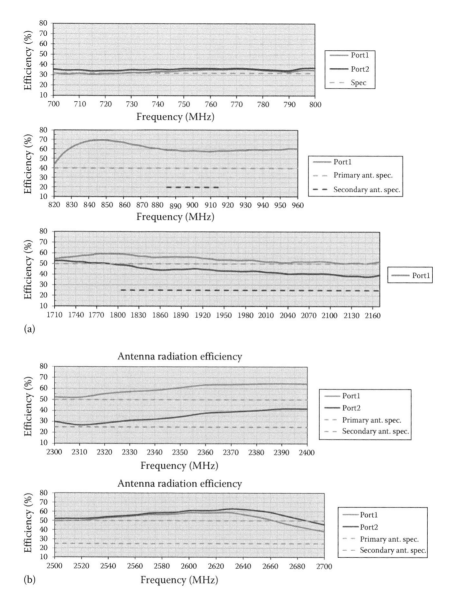

FIGURE 4.10 (a) Radiation efficiency plots for tuned-match antenna, low and mid bands and (b) radiation efficiency plots for tuned-match antenna, high bands.

4.8 CONCLUSIONS

Antenna design continues to be challenged by the increasing number of bands, the requirement for multiple antennas to support MIMO operation, as well as the thinner form factors so popular with latest model smartphones and other wireless devices. With further complications arising from intermodulation and harmonic radiation requirements, and the promise of LTE-Advanced, it is likely that antenna tuning will find application in an increasing number of commercially produced wireless devices as tuning products mature and antenna designs continue to mature.

REFERENCES

1. Wheeler, H., Small antennas, *IEEE A&P*, AP-23(4), 462–469, 1975.
2. Chu, L.J., Physical limitations of omni-directional antennas, *Journal of Applied Physics*, 19, 1163–1175, 1948.
3. Harrington, R., Effect of antenna size on gain, bandwidth, and efficiency, *Journal of Research of the National Bureau of Standards, Section D: Radio Propagation*, 64D(1), 1–12, 1960.
4. Reading, L.J., Designing dual-band internal antennas, *Electronic Design News*, 99–104, November 8, 2001.
5. McClean, J.S., A re-examination of the fundamental limits on the radiation Q of electrically small antennas, *IEEE A&P*, 44, 672–675, May 1966.
6. Grimes, D. and Grimes, C.A., Minimum Q of electrically small antennas: A critical review, *Microwave and Optical Technology Letters*, 28(3), 172–177, February 5, 2001.
7. http://www.ctia.org/content/index.cfm/AID/10021.
8. http://www.3gpp.org/IMG/pdf/2009_10_3gpp_IMT.pdf.
9. Parson, S.D., *The Mobile Radio Propagation Channel*, 2nd edn., Wiley, London, UK, 2000.
10. Breit, G. and Ozaki, E., Phone level radiated test methodologies for multi-mode multi-band systems, *IWPC China Mobile Workshop, IWPC Transactions Handset Antenna Technologies for MultiMode-Multi-Band*, Durham, NC, January 2007.
11. Fabres, M.C., Systematic design of antennas using the theory of characteristic modes, PhD thesis, Universidad Polytecnia de Valencia, Valencia, Spain, February 2007. http://anteny.jeziorski.info/wp-content/uploads/2011/10/2007_Fabres.pdfs, accessed March, 2015.
12. Blanch, S., Romeu, J., and Corbella, I., Exact representation of antenna system diversity performance from input parameter description, *Electronics Letters*, 39(9), 705–707, May 1, 2003.
13. Kalliola, K., Sulonen, K., Laitinen, H., Kivekas, O., Krogerus, J., and Vainikainen, P., Angular power distribution and mean effective gain of mobile antenna in different propagation environments, *IEEE Transactions on Vehicular Technology*, 51(5), 823–838, 2002.
14. Long, B. et al., A simple broadband dipole equivalent circuit model, *Proceedings of the IEEE International Symposium on Antennas and Propagation*, vol. 2, Salt Lake City, UT, July 16–21, pp. 1046–1049, 2000.
15. Ramo, S., Whinnery, J.R., and Van Duzer, T., *Fields and Waves in Communication Electronics*, Wiley, New York, 1965, Section 11.13.
16. Oh, S.-H. et al., Automatic antenna-tuning unit for software-defined and cognitive radio, *Wireless Communications and Mobile Computing*, Wiley Interscience, 2007. Published online in Wiley InterScience (www.interscience.wiley.com) DOI: 10.1002/wcm.484.
17. Kundert, K., Accurate and rapid measurement of IP2 and IP3, Designers Guide Consulting, May 22, 2002. http://www.designers-guide.org/analysis/intercept-point.pdf, accessed March 2015.
18. Caimi, F.M. and Montgomery, M., Dual feed, single element antenna for WiMAX MIMO application, *IJAP*, 2008, Article ID 219838, April 2008.

5 Tunable and Adaptive Antenna Systems

Laurent Desclos, Sebastian Rowson, and Jeff Shamblin

CONTENTS

5.1 PASSIVE ANTENNA LIMITATIONS

5.1.1 COMMON PASSIVE ANTENNA APPROACHES IN WIRELESS DEVICES

A wide variety of antenna types are utilized in wireless communication devices. The variation in the number of frequency bands needed to service along with the devices being designed for either cellular or ISM frequency bands is one of factors that determines antenna type selection. MIMO systems are gaining favor in WLAN as well as cellular applications, which brings isolation and envelope correlation coefficient (ECC) requirements into the mix. Considering that most antennas in current commercial wireless devices are internal, it can be challenging to meet the electrical requirements levied on the antenna system.

Basic antenna types found in wireless devices are monopole elements, inverted F antennas (IFAs), planar inverted F antennas (PIFAs), and loops. Each antenna type has advantages and disadvantages that need to be considered as trade-offs are made to balance frequency bandwidth with performance. For example, monopole elements tend to couple strongly to the surrounding structure in the host wireless device, which will reduce isolation between adjacent antennas; on the other hand, the monopole element will typically possess wider frequency bandwidth compared to loops and PIFAs. PIFAs are 3D structures where multiple resonances can be generated to allow for multiband operation, but like monopoles tend to couple strongly to the host device. Loops do not couple well to the surrounding structure of a host device, making this type element better suited for embedded applications where isolation is required. Unfortunately, the loop is a narrow bandwidth element, which is not suited for multiband applications.

More recently, new antenna structures or topologies have been developed to address the need to maintain isolation between an embedded antenna and its surroundings while maintaining bandwidth and efficiency. An example of a well-isolated antenna is the isolated magnetic dipole (IMD). This type of antenna is designed where consideration is made as to the near-field characteristics of the antenna to minimize coupling between the antenna and surroundings, resulting in improved isolation between antennas in a device, which is needed as more transceivers are integrated into wireless devices and as MIMO systems gain favor in these communication systems.

5.1.2 LIMITATIONS IN BANDWIDTH

Modern wireless devices, such as smartphones, tablets, laptops, and wearable devices tend to have small form factors and are thin to increase usability (e.g., smartphone that fits in a shirt pocket). Designing a lot of functionality into a smartphone (cellular, WLAN, and GPS receivers and transceivers) highlights the need to reduce or limit the volume that an antenna can occupy. Not having adequate volume reduces the performance of the antenna, with the bandwidth at the lower frequencies being the first parameter where degradation is observed.

The main problem that arises when limiting the size or volume of an antenna is achieving sufficient operational bandwidth. Wheeler[1] defined the following general formula that links the bandwidth of an antenna to its mode volume at a certain frequency:

$$\frac{\Delta f}{f} = K \times \frac{\text{antenna mode volume}}{(\text{radio wavelength})^3}$$

This equation shows that the bandwidth Δf over the central frequency f is linked by a dimensionless number K to the ratio of the antenna mode volume to the wavelength. The K factor is a figure of importance when, all things being equal, we want to compare one antenna to another. The K factor is related to the antenna technology and how it is designed. In Wheeler's original paper, the antennas used to demonstrate this were electric or magnetic dipoles and loops in freespace. This allowed the overall antenna mode volume to be defined according to its natural boundaries. For the dipole, it was the sphere enclosing the dipole. In the case of electrically small antennas, the problem is defining the antenna mode volume since in most cases it significantly exceeds the physical volume of the antenna itself. Indeed, some antennas

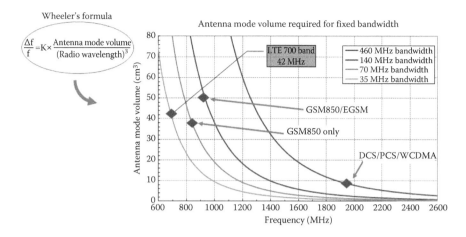

FIGURE 5.1 Antenna mode volume related to frequency of operation and bandwidth.

tend to couple strongly to the ground plane making the whole device the antenna. This might lead to an apparent advantage when bandwidth alone is considered but has many drawbacks for system performance and multiple antenna integration. For this reason, the antenna alone should be designed to provide sufficient bandwidth. Figure 5.1 shows a graph relating antenna mode to various center frequencies. The curves shown are curves of fixed bandwidth, varying from 35 to 460 MHz. As can be seen from this set of curves, as the center frequency of operation decreases and as the required bandwidth increases, the antenna mode volume needed increases. An important point to take away from this graph is that it requires a much smaller volume to service 460 MHz of bandwidth at 2 GHz compared to 700 MHz.

5.1.3 LIMITATIONS DUE TO THE CHANGING ENVIRONMENT

A wide variety of wireless devices, such as cell phones, tablets, and smart watches, are required to operate across a wide range of use cases, with these use cases introducing a detuning effect on the antenna system. For example, a cell phone is required to maintain a certain level of performance across use cases that include phone in user's hand, phone against the user's head while in hand, and phone placed on a surface such as a table. The cell phone against the user's head will typically introduce a frequency shift in the antenna as well as decrease the total efficiency of the antenna due to mismatch loss and absorption loss. A passive antenna for a cell phone will need to operate across these multiple use cases, so the antenna cannot be optimized for any one use case without degrading performance across the other use cases.

5.1.4 A NEED TO OPTIMIZE RADIATION PATTERNS TO IMPROVE THE COMMUNICATION LINK

Today's wireless device, such as a smartphone or WLAN-enabled tablet or laptop, is consistently operating in a multipath environment. The multiple reflections due to walls, furniture, and other obstructions cause fading in the propagation channel. Over the years, antenna diversity schemes have been implemented in cell phones and WLAN access points to compensate for deep fades, with these antenna diversity schemes incurring volume, complexity, and cost constraints during product development. MIMO systems, on the other hand, rely on a rich multipath environment to assist in signal discrimination during the processing phase of the coincident transmit and receive signals. Though multipath is desired for good MIMO system performance, the fixed radiation pattern of the antennas that constitute the MIMO system limit the performance of the wireless device, specifically in terms of MIMO system coverage across all aspect angles. For both antenna diversity and MIMO applications, the inability to dynamically alter the radiation pattern of the embedded antennas limits the performance of the antenna system, primarily due to the fact that the gain maxima of the antenna is typically not optimized for mobile systems due to the motion on the mobile side.

5.2 TUNABLE MATCHING CIRCUITS

The limitations of passive antennas previously highlighted point toward a need to dynamically adjust antenna parameters such as impedance properties. By dynamically adjusting the impedance match of an antenna, the antenna can be optimized as the various use cases of a wireless device are exercised. A tunable matching circuit can be formed at the feedpoint of an antenna, with this matching circuit commonly consisting of one or two tunable capacitors and some passive components, such as capacitors and inductors. Various circuit topologies can be configured and both open-loop and closed-loop matching can be utilized to dynamically match the antenna. Considering that most commercial wireless devices have small form factor embedded antennas, a revisit of Figure 5.1 can provide insight as to what portion of the frequency range will benefit most from tuning. Figure 5.1 shows that 70 MHz of bandwidth centered at 1900 MHz can be supported with less than 2 cm³ of antenna mode volume. Moving down in frequency, 70 MHz of bandwidth centered at 854 MHz requires 38 cm³ of antenna mode volume. This trend indicates that for a dual resonance antenna, which has a low-frequency resonance and a high-frequency resonance, antenna tuning will be required at the low-frequency resonance when the antenna impedance is being tuned to increase the bandwidth. For Long Term Evolution (LTE) cellular applications, carrier aggregation is a feature that is being implemented where channels in multiple frequency bands are used simultaneously to transmit and receive data. A set of frequency band pairings have been defined in the 3GPP standard, with the end result being that when tuning is applied to the antenna in a wireless mobile device, it is best to tune the antenna at the low-frequency band since this is the frequency band that will require the most improvement; a wide bandwidth response at the high-frequency resonance will allow for multiple carrier aggregation band pairings.

The topology of the tuning circuit is an important consideration, with the main consideration to be made being whether to implement a series or shunt tunable capacitor. A series configuration for the tunable capacitor will result in more stringent Q and power handling requirements placed on the tunable capacitor due to the higher current flowing through the capacitor. To maintain low-loss tuning circuit character-istics, a series configuration will require a Q in the range of 30–80, with this range in Q dependent on the frequency of operation and antenna type. A shunt capacitor configuration will allow for relaxed Q and power handling requirements compared to a series configuration, which can translate to a reduction in tuning losses and reduced component cost. Qs in the range of 10–30 can support low-loss tuning cir-cuit requirements when the antenna design is taken into consideration and the tuning circuit and antenna are designed as a system. A proper selection of antenna type can allow for the use of a shunt tunable capacitor, which will provide the impedance tun-ing range required to impedance match the antenna for the host device design and use cases. To demonstrate the effect of Q on shunt capacitor performance in a tuning circuit, a set of circuit simulations were performed where Q was varied and insertion loss and return loss were monitored. A shunt capacitor was positioned between two 50 ohm ports with 50 ohm transmission lines connecting the two 50 ohm ports to the terminals of the capacitor. The Q of the capacitor was varied from 10 to 100 and the return loss and insertion loss of the circuit were monitored. Figure 5.2 shows the circuit that was simulated along with the insertion loss and return loss as the Q of the capacitor is varied. As can be seen at 820 MHz, the insertion loss of the circuit varies by less than 0.15 dB when the Q is varied from 10 to 100. The return loss varies by less than 0.2 dB across the 500 MHz to 10 GHz range.

5.2.1 Adaptive Matching

Open- and closed-loop impedance matching have both pros and cons in terms of capabilities and system integration. Closed-loop impedance matching provides the capability of impedance matching an antenna without input or control signaling from the rest of the system (transceiver, baseband), which minimizes the mismatch loss as the environment changes. Closed-loop matching does need to synchronize with the transmit and receive functions and requires additional RF circuitry for implementa-tion, which incurs cost and losses. Open-loop impedance matching is easier to imple-ment, but typically requires input from the communication system such as frequency band or channel to optimize for. A third technique discussed here, adaptive match-ing, which can be considered a hybrid technique where additional capability can be gained compared to an open-loop matching scheme by using additional information from a baseband to provide additional tuning capability.

Current commercial wireless devices have one or multiple sensors or devices that can be used to discern the use case or environment that the antenna system in a wire-less device is exposed to. These sensors can be proximity sensors, display status, speaker state, microphone status, and/or orientation sensor. By taking an antenna element, tunable matching circuit formed with a tunable capacitor and passive com-ponents, and a lookup table relating sensor inputs to capacitor settings, an adaptive antenna solution can be implemented in a wireless device that matches the antenna for a specific frequency band and adjusts the impedance match of the antenna to compensate for an estimated use case or environment. Figure 5.3 shows a sche-matic of an adaptive matched antenna, with baseband of the host device providing information such as frequency band or channel and other stimuli for use in making decisions regarding antenna tuning. Figure 5.4a shows a plot of return loss for an adaptive antenna embedded in a cell phone; in this plot, there are several traces related to specific tunable capacitor tuning states. This return loss plot shows three resonances: low, mid, and high band, which provides the capability to cover a majority of 3G and 4G cellular bands with a single port antenna, which is typical for main antennas in current cell phones. The low-band resonance can be seen to shift in

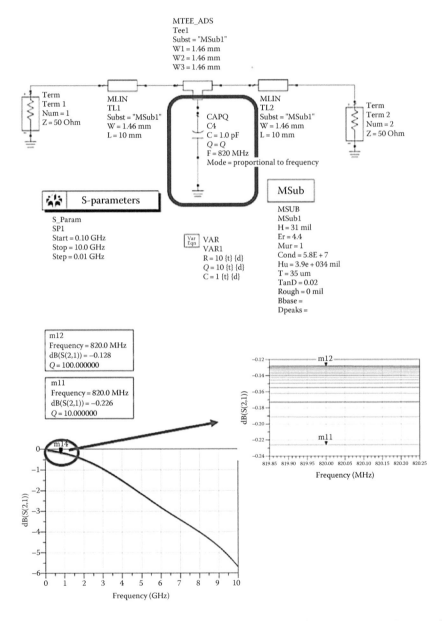

FIGURE 5.2   Circuit simulation where Q of a shunt capacitor is varied.   (*Continued*)

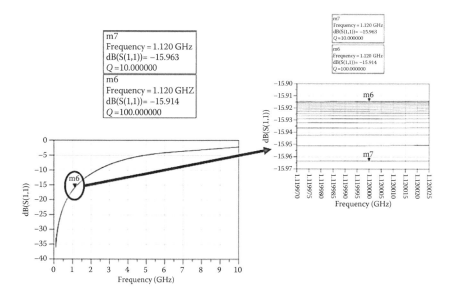

FIGURE 5.2 (*Continued*)   Circuit simulation where Q of a shunt capacitor is varied.

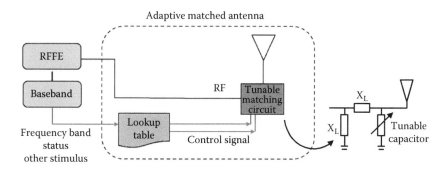

FIGURE 5.3   Schematic of an adaptive matched antenna.

frequency as the tunable capacitor is varied, with the mid- and high-band resonances either being optimized or detuned as the tunable capacitor is varied. The return loss traces shown here are for a "freespace" use case defined as a use case where there are no objects in the vicinity of the cell phone (neither hand nor head loading occurs for this use case). Figure 5.4b highlights the benefit of an adaptive matched antenna. Instead of the wideband frequency sweep shown in Figure 5.4a,b shows the return loss of the low-band resonance for the hand-loading use case. The multiple return loss traces relate to specific capacitor tuning states. The desired frequency band of interest is denoted with the markers at 824 and 894 MHz, with the resonances to the

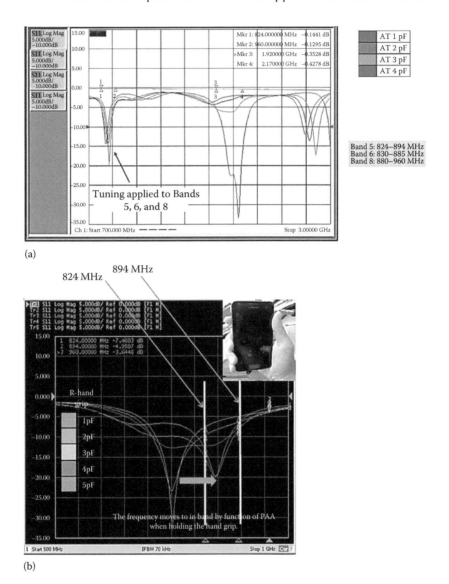

(a)

(b)

FIGURE 5.4 (a) Return loss showing three resonances from the single-feed antenna installed in a cell phone; multiple traces relate to different capacitor tuning values. (b) Return loss corrected for hand loading by adjusting the value of the tunable capacitor in the matching circuit.

left of these markers representing the return loss of the antenna with hand loading of the cell phone prior to selecting the optimal capacitor tuning state. As the tunable capacitor is varied, the return loss can be improved at the frequency band of interest. Proximity sensor status is used to determine which capacitor tuning state to command, with the lookup table relating such functions as frequency band, proximity sensor status, speaker, or microphone status to tunable capacitor tuning states. The lookup table is developed during the adaptive antenna design phase.

5.3 BAND SWITCHING TECHNIQUES

A second method of dynamic tuning that can be applied to an antenna is termed *band switching* and is also referred to as an *active aperture*. This technique differs from the impedance-matching techniques described previously in multiple respects: instead of tuning the impedance at the feedpoint of the antenna, the tuning implemented in band switching is applied at the radiating portion of the antenna. Instead of being concerned with altering the impedance match of the antenna, band switching is implemented to shift the frequency response of the antenna. This technique is useful to implement when a volume constrained antenna is required to cover a frequency range that exceeds the bandwidth performance that can be achieved from the passive antenna structure. A well-implemented band switching design will increase the bandwidth of the antenna by shifting the resonant frequency of the radiating portion of the antenna, while an open- or closed-loop impedance matching scheme will attempt to improve the impedance match of the radiator without addressing the need to shift the resonance of the radiator.

5.3.1 Approaches to Alter the Frequency Response of a Radiator

Multiple approaches can be implemented to dynamically shift the frequency response of a radiator. Each approach requires that a tunable component, such as a switch, tunable capacitor, or diode by connected or coupled to the radiating element in a fashion that affects the electrical length and/or current mode on the radiator. Two approaches are described here, with the approaches different in respect to whether both low- and high-band resonances of the antenna shift in frequency or one resonance only. A dual-resonance antenna is considered here due to the prevalent use of dual-resonance antennas to provide antenna capabilities at the various 3G and 4G LTE frequency bands commonly used in cell phone applications. The second approach described here provides the capability to satisfy a majority of LTE carrier aggregation band pairings between lower frequency bands and upper frequency bands.

5.3.1.1 Band Switched Antenna Configuration 1

A first approach works well with IMD-type antennas as well as PIFAs when these antenna types are embedded internally to wireless devices, such as cell phones and laptops. A 3D radiator is formed from a conductor that is positioned above a ground plane, with the circuit board of the host wireless device acting as the ground plane. A feed leg is used to connect the planar portion of the antenna elevated above the ground plane to the transceiver, and a ground leg is typically included in the design to aid in matching the antenna and in setting up the desired current mode. A parasitic element is positioned between the bottom side of the antenna element and the ground plane, and an RF switch is connected to the common port of the RF switch. One port of the RF switch is configured in an "open" state, where the port is not connected to any structure or component. The additional ports of the RF switch are connected to the ground either directly or using lumped components, such as capacitors and inductors. The frequency response of the antenna can be shifted by switching from one port of the RF switch to another port. More specifically, when the open switch port is activated, the parasitic element has negligible effect on the frequency response of the antenna. This is due to the negligible level of coupling between the antenna and the parasitic and the electrical length of the parasitic element. When a second switch port, connected to a reactive lumped component and the component in turn connected to the ground is activated, the frequency response of the antenna will shift. The shift in frequency response is due to two effects: (1) the electrical length of the parasitic element is altered when the reactive lumped component is connected to it, and (2) the level of coupling between the parasitic element and the antenna element is increased when the parasitic element is attached to the ground (through a lumped reactance, in this case) (Figure 5.5). Figure 5.6 shows the antenna structure with parasitic element for two reactive loading states that provide different frequency responses, with the first loading state of the parasitic element being an open circuit and the second loading state being a fixed value capacitor. Figure 5.7

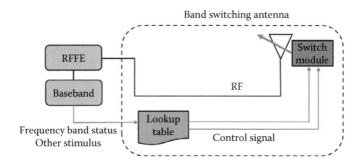

FIGURE 5.5 Schematic of a band switched antenna.

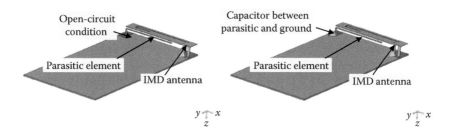

FIGURE 5.6 IMD antenna configuration 1 with parasitic element; two reactive loading states are shown.

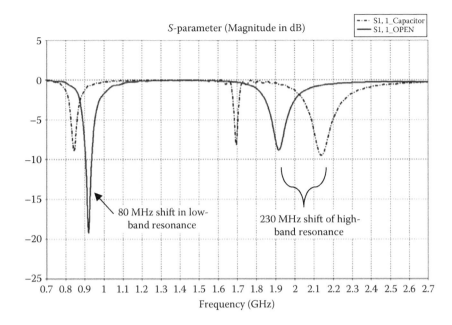

FIGURE 5.7 Return loss of the two loading states of the parasitic element for antenna configuration 1 shown in Figure 5.6.

shows the return loss for both loading states of the IMD antenna; notice that both low- and high-band resonances have shifted when the reactive loading on the parasitic element changes from an open circuit to a capacitive reactance. The lower frequency response shifts lower in frequency while the upper frequency response shifts higher in frequency.

5.3.1.2 Band Switched Antenna Configuration 2

A second band switched antenna configuration is described using the same concept as discussed in configuration 1 where a parasitic element is positioned beneath an IMD antenna. The IMD antenna in this second configuration is different in design, with a large portion of the conductor positioned orthogonal to the ground plane, with this portion of the conductor used to provide a wide bandwidth upper frequency resonance. The parasitic element is positioned beneath the IMD antenna to affect the low-band resonance while providing negligible coupling to regions of the antenna element that effect high-band performance. With this configuration, a change in reactive loading of the parasitic element will result in the low-band resonance shifting in frequency while maintaining a near constant high-band resonance. As previously mentioned, this antenna configuration is good for antenna designs where 4G LTE carrier aggregation requirements are levied. Figure 5.8 shows the antenna structure with parasitic element for two reactive loading states that provide different frequency responses, with the first loading state of the

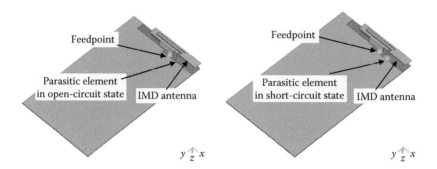

FIGURE 5.8   IMD antenna configuration 2 with the parasitic element; two reactive loading states are shown.

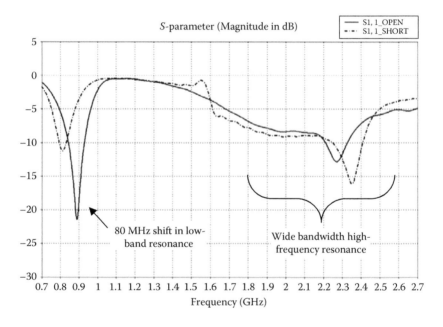

FIGURE 5.9   Return loss of the two loading states of the parasitic element for antenna configuration 2 shown in Figure 5.8.

parasitic element being an open circuit and the second loading state being a fixed value capacitor. Figure 5.9 shows the return loss for both loading states of the IMD antenna; notice that in this configuration, the both low resonance has shifted when the reactive loading on the parasitic element changes from an open circuit to a capacitive reactance while the high-band frequency response remains practically unchanged.

5.3.2 COMBINATION OF ADAPTIVE MATCHING AND BAND SWITCHING TO OPTIMIZE AN ANTENNA

The adaptive matching and band switching techniques can be combined in a single antenna design to further optimize the impedance and resonant frequency characteristics. One benefit to highlight in this combined approach is the natural filtering that the antenna can provide to improve the transceiver performance and to reduce some filtering requirements. The band switching technique in the form of a multiport switch coupled to the radiator provides the capability to frequency shift, a narrow bandwidth frequency response to a desired center frequency while maintaining the narrow bandwidth properties of the antenna, which can be used to provide filtering attributes. The adaptive matching technique in the form of a tunable capacitor at the feedpoint of the antenna can then be applied to fine tune the impedance properties and retune the antenna as the host device encounters different use cases (e.g., a cell phone transitioning from hand use case to head and hand use case). Figure 5.10 shows a

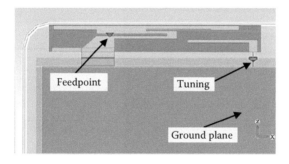

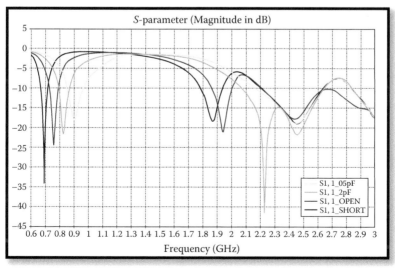

FIGURE 5.10 Antenna design, return loss, and total efficiency of a band switched antenna showing frequency response roll-off at low-band resonances.   (*Continued*)

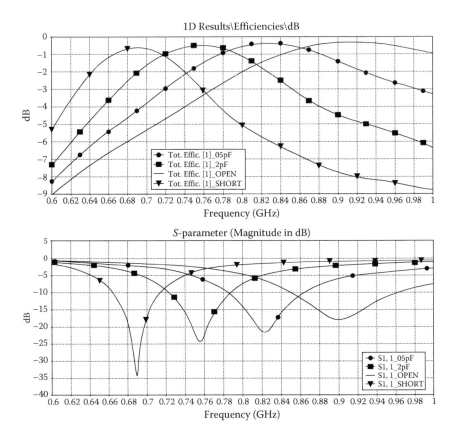

FIGURE 5.10 (*Continued*) Antenna design, return loss, and total efficiency of a band switched antenna showing frequency response roll-off at low-band resonances.

band switching antenna designed for integration into the edge of a host device where four low-band tuning states are set up to provide tuning across the 700–960 MHz frequency range. The roll-off in return loss and total efficiency shows that several dB of attenuation can be achieved out-of-band for this type of antenna compared to the broader bandwidth passive antennas, with this additional attenuation providing better isolation.

5.4 DYNAMIC RADIATION PATTERN OPTIMIZATION

Previously, we described an adaptive matched antenna, where the impedance of the antenna is dynamically adjusted to improve the match of the antenna port to the rest of the communication system. This technique will reduce the mismatch loss of the antenna and better match the antenna to the power amplifier. Another active antenna technique that can provide substantial benefit in terms of increased antenna system efficiency and system level effects such as increased throughput is dynamic radiation pattern optimization. By developing a technique to dynamically adjust the radiation pattern of an antenna, the antenna can now be optimized for specific multipath environments. This technique of dynamically changing the radiation pattern of an antenna can be thought of as a method to match or optimize the second port of an antenna, the "radiating port," to the propagation channel of a communication system. This dynamic adjustment of the radiation pattern of the antenna can be implemented by adjusting the radiation pattern of a single embedded antenna, termed a "modal antenna" or can be implemented by using traditional antenna array techniques where two or more antenna elements are combined to generate a directive pattern that can be scanned by adjusting the phasing of the feed network.

5.4.1 Description of the Modal Antenna Technique

Ethertronics Inc., an antenna system company, has devoted several years to the development of modal antenna concepts and techniques, with this development culminating in the ability to dynamically adjust the radiation pattern of a single port small form factor antenna. This modal antenna technique is based on the use of an IMD antenna as the basic element, and additional elements or components are implemented to alter the current mode on the IMD antenna. The modal antenna technique is described here, and the term Null steering refers to the dynamically changing the radiation pattern of the modal antenna, resulting in the capability of shifting the nulls in the radiation pattern.

The IMD antenna in the fundamental mode exhibits a radiation pattern similar to a dipole antenna structure with well-defined peaks and nulls in the 3D radiation patterns. On closer inspection, it was seen that if the nulls in pattern could be rotated by $90°$ (from $-X$ to $-Z$ and from $+X$ to $+Z$) then it could be possible to have a peak at the two locations ($+Z$ and $-Z$), which were previously occupied by nulls in the pattern. A composite plot of the two patterns would provide coverage along $+X$, $+Z$, $-X$, and $-Z$ directions. This can potentially account for several dB of improvement in null coverage. The current distribution characteristics of the IMD antenna were exploited further to develop a structure that could steer the nulls in the radiation pattern of the IMD antenna structure. This was achieved using a parasitic element placed offset from the main IMD antenna. This parasitic element was switched between an open-circuit and short-circuit state. When the parasitic element is open circuited, the antenna structure generates a resonance at frequency f_0; when the parasitic element is shorted to the same ground plane as the antenna, a dual mode response (f_1 and f_2) is generated. It can be seen from the current distribution plots in Figure 5.11 that the resonance at frequency f_1 is generated from the IMD antenna and the second resonance f_2 is generated from the parasitic element. On further inspection, it is found that the current distribution characteristics and also the radiation patterns are identical for frequency point f_0 and f_1. In order to better describe the concept, a simulation model for a basic null steering antenna is shown in Figure 5.12.

The antenna design involves capacitive coupling between a parasitic element and a main drive element to dynamically alter the surface current distribution on the structure and as a result alter the far-field radiation patterns. The different radiation states generated by reactively loading the parasitic element are referred to as *Modes*. It should be noted that the term mode here does not correspond to the transverse electric (TE)

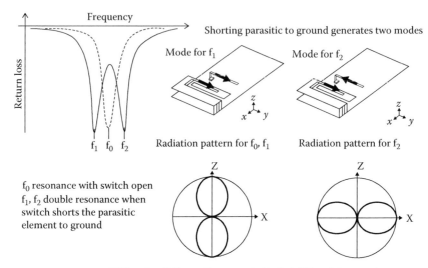

FIGURE 5.11   Basic principle of null steering an IMD antenna.

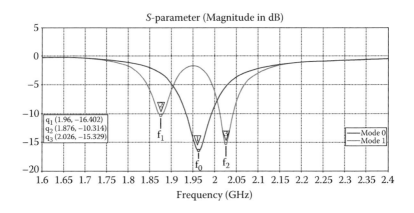

FIGURE 5.12   Simulation model of a null steering antenna.

or magnetic (TM) modes and is used simply as a term to specify the radiation state of the antenna. The antenna structure generates two fundamental radiation states Mode 0 and Mode 1 associated to when the parasitic is in an open-circuit state and short-circuit state, respectively. Mode 0 exhibits a single frequency response centered at frequency f_0 generated from the IMD antenna element. Mode 1 exhibits a dual frequency response at frequencies f_1 and f_2, wherein f_1 exhibits a radiation pattern similar to the IMD antenna element (f_0), and f_2 exhibits a radiation pattern wherein the nulls in the radiation pattern for f_1 are rotated by almost 90° as seen in Figure 5.14.

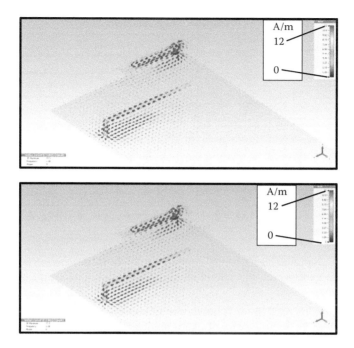

FIGURE 5.13   Current distribution plots at frequencies f_1 and f_2 of a null steering antenna.

The surface current distribution plots at frequencies f_1 and f_2 in Figure 5.13 show that for f_1, the current distribution vector on the parasitic arm is in the same direction as the inner arm of the IMD antenna element, whereas it is in the opposite direction compared to the inner arm of the IMD antenna in the case of frequency f_2. This variation in the near-field characteristics of the antenna structure result in a variation in the far-field characteristics at frequencies f_1 and f_2. The far-field variation can be measured in terms of change in the co- and cross-polarization gain between frequencies f_1 and f_2. The 3D radiation patterns for f_1 and f_2 in Figure 5.14 also show that the

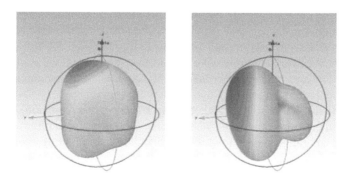

FIGURE 5.14   3D radiation pattern at frequencies f_1 and f_2 for a null steering antenna.

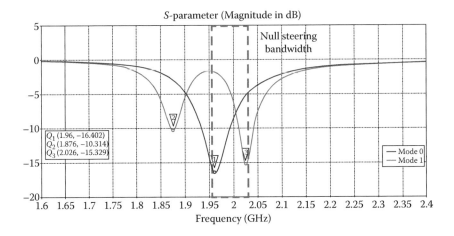

FIGURE 5.15   Null steering bandwidth.

null in the radiation pattern for f_2 (along the +Z axis) is rotated by almost 90° in the radiation pattern for f_1 (along the –X axis) (Figure 5.15).

The parasitic element placed offset from the main IMD antenna element can be terminated in either open-circuit or short-circuit state to generate to fundamental radiation states at frequencies f_0 (Mode 0) and frequency f_2 (Mode 1). These two frequency points define the frequency bandwidth over which it is possible to generate two or more radiation patterns, which will be de-correlated in far field. In terms of design guideline, these two states will be the fundamental states and it is possible to generate more states by terminating the offset parasitic elements in lumped capacitors, inductors or LC circuits.

This modal antenna technique provides the ability to generate multiple radiation patterns from a single port antenna, and a direct application of this type antenna is for a receive diversity in 3G handset applications. A single modal antenna can be implemented in a 3G application where the receive diversity antenna and second receive port in the transceiver can be eliminated. Using a modal antenna as the main antenna in the handset will provide switched diversity from the single modal antenna. This implementation will result in reduced power consumption due to the elimination of the second receive port in the transceiver, and will also result in reduced cost and reduced internal volume required from the antenna system. The modal antenna can address these issues by proving pattern diversity from a single radiating structure without adding a second receive chain. The modal antenna architecture is being designed keeping these aspects in mind with an aim of providing an antenna system which provides the maximum diversity gain and improves the link budget by adding several dB of improvement to the fading margin. In essence, the modal technique utilizes novel antenna architecture capable of generating two or more unique modes with low ECC and comparable efficiencies. The technique will utilize a switched combining technique to switch between the

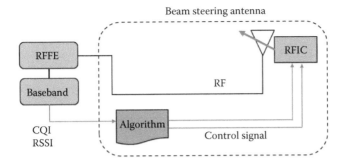

FIGURE 5.16 Model antenna system in a communication device.

modes. The goal here would be to have possible gains that would be comparable to maximum ratio combining (MRC), which increases the hardware required and the associated complexity.

The antenna hardware for the modal antenna is satisfied with an IMD antenna, a parasitic element, and multiport RF switch. To complete the modal antenna system, an algorithm is developed and implemented to provide the decision-making process for optimal radiation mode selection in a dynamic environment. The algorithm requires a single metric from the host device baseband processor, with this metric being a low-latency SINR, RSSI, CQI, or equivalent metric, which characterizes the quality of the communication link. The algorithm can be located in any processor in the host device such as baseband or the application processor. The role of the algorithm is to compare the link quality metric to the sampled radiation modes and choose the optimal mode for the communication link. A periodic sampling of the additional modes (the modes not used for communication at a specific instance) is made to determine, which mode to switch to and when to switch to keep the communication link optimized. Figure 5.16 shows a schematic of a modal antenna system implemented in a communication device.

The modal antenna architecture would be designed keeping these aspects in mind with an aim of providing an antenna system, which provides the maximum diversity gain and improves the link budget by adding several dB of improvement to the fading margin.

5.5 CONCLUSIONS

The adaptive antenna techniques described in this chapter, adaptive matching, band switching, and beam steering, provide the capability of dynamically altering antenna characteristics, such as impedance, resonant frequency, and radiation patterns. These adaptive antenna techniques are realized by taking advantage of onboard sensors and signaling found in today's commercial wireless devices. These sensor inputs can be proximity sensor status, frequency band information, display lighting status, microphone, speaker state, and so on. One or multiple adaptive antenna techniques can be combined to optimize multiple parameters simultaneously.

For high production volume applications, a customized **RFIC** containing switches, tunable capacitors, and a processor (e.g., to house a beam steering algorithm) can be developed to bring these adaptive antenna techniques to a single antenna structure is a cost-effective manner.

REFERENCE

1. Wheeler, H.A., Fundamental limitations of small antennas, *Proceedings of the IEEE*, 35(12), 1479–1484, December 1947.

6 Effective Antenna Aperture Tuning with RF-MEMS

Larry Morrell and Paul Tornatta

CONTENTS

6.1 ANTENNAS AND MEMS: ADVANTAGES OF MEMS FOR APERTURE-TUNED ANTENNAS

6.1.1 KEY PERFORMANCE PARAMETERS OF MEMS

Several trends in the phone market are increasing the difficulty of designing antennas for mobile devices. These include wider frequency coverage from 700 MHz to 3.6 GHz for Long Term Evolution (LTE) services and shrinking size allocated to the antenna (remember the telescoping antennas of yesteryear?). The industrial design of smartphones, where the screen takes up nearly all the area and sometimes the case is made of metal—effectively blocking all radio frequency (RF) signals causes further difficulty. Thus far, approaches to solving the problem have fallen short of ideal.

One approach is to settle for lower efficiency. Broadband antennas have worse efficiency than narrowband antennas for a given antenna size. But if lower efficiency is acceptable for a particular device design, a broadband antenna with sufficient band coverage will work. The trade-off is that the mobile device's poor efficiency will mean smaller coverage area, more frequently dropped calls, lower data rate, and shorter battery life.

Another approach is to design an antenna with narrow instantaneous bandwidth and use a matching network to optimize the power transfer across the terminals of the antenna. This technique offers some benefit to the RF chain, allowing the RF amplifier circuits to operate more effectively. However, this technique does not fundamentally change the radiating element resonant frequency and as the matching network moves the frequency farther from resonance, the radiation efficiency drops off.

The best approach to the antenna problem that overcomes the limitations of either of these previous approaches, is to incorporate the tuner directly in the aperture of the antenna, so when the tuner is actuated, the resonant frequency of the radiating structure is changed. This is called *aperture tuning*. If the antenna is matched to 50 ohms at resonance, changing the resonant frequency of the structure will maintain the impedance match. Aperture tuning accomplishes two things: (1) it maximizes the radiation efficiency of the antenna, and (2) it maintains the antenna impedance match to optimize the power transfer across the antenna terminals. Antenna aperture tuning allows antennas to be designed with narrow instantaneous bandwidth that can be tuned to different frequency bands when needed.

Aperture tuning is not a new idea. There are examples of aperture-tuned antennas in ham radios, land mobile radios, and military applications. The tuning devices used in these applications have very high performance with Quality factor $(Q) > 100$, equivalent series resistance $(ESR) < 0.5$ ohm, voltage handling capability > 40 V RMS, and high linearity $IP3 > 70$ dBm. These high-performance devices, however, do not meet the additional requirements for consumer electronics products that are small in size, low in cost, and high in reliability (a large number of tuning cycles).

Only recently have microelectromechanical systems (MEMS) emerged as a technology that addresses all of the technical and commercial requirements for antenna aperture tuning.

6.1.2   TRADE SPACE OF MEMS IMPLEMENTATIONS

One of the most important features of RF MEMS is high Quality Factor (Q) across all frequencies. The main reason why RF MEMS can have such high Q is that the basic structure is a metallic, electromechanical switch, not a semiconductor transistor acting as a switch. The performance is limited by the resistivity of the metals (R), the parasitic inductance (L), and capacitance (C). The following equation shows the relationship between R, L, and C in determining Q.

$$Q = \frac{1}{(1/Q_L) + (1/Q_C)}, \quad \text{where} \quad Q_c = \frac{1}{\omega C R_C} \quad \text{and} \quad Q_L = \frac{\omega L}{R_L} \tag{6.1}$$

Keeping the resistance low and minimizing parasitic inductance and capacitance allows the MEMS device to approach the performance of an ideal component.

For antenna aperture tuning, all of the key performance parameters must be optimized simultaneously. RF MEMS allows for the optimization of all parameters without having to trade-off between them.

6.1.3   ANTENNA SELECTION FOR ANTENNA APERTURE TUNING

Aperture tuning can be applied to a wide variety of antenna types. The examples will focus on antenna types that are applicable for mobile handset and smartphone devices. These antenna types are typically planar inverted F antennas (PIFA), monopole, loop, and slot antennas. All of these antennas can be modeled as distributed networks of R's, L's, and C's. The resonant frequency of the antenna can be changed by changing either the capacitance or inductance at different points in the distributed network. In a physical antenna, the location of the tuning element or variable reactance will determine how much capacitance or inductance it takes to shift the resonant frequency. The location also determines the voltage across the terminals of the tuner and as a consequence, the linearity of the device. In general, regions of high voltage will allow for greater tuning range with small changes in capacitance but will require higher linearity to tolerate the voltage.

Specific antenna selection for the product is a combination of design experience, industrial design constraints, and frequency band requirements.

6.2   TRADE-OFF OF INSTANTANEOUS BANDWIDTH AND SYSTEM REQUIREMENTS

6.2.1   SYSTEM OPERATION: HOW WIDE DOES THE BANDWIDTH NEED TO BE AT ANY ONE TIME

Modern smartphones are required to cover a large number of frequency bands and protocols (Figure 6.1).

Under normal operation, the phone is required to operate in only one band, while keeping the ability to quickly look at other frequency bands and protocols

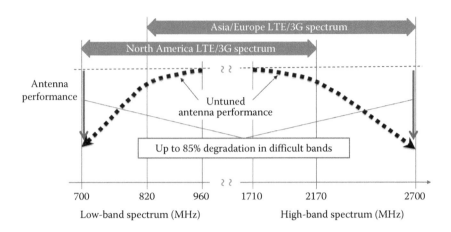

FIGURE 6.1   Span of cellular frequencies.

FIGURE 6.2   Carrier aggregation.

for roaming, hand-off, and other system management functions. Carrier aggregation (CA) requires the phone to operate in two bands at the same time. CA requirements depend on the wireless service provider and regional requirements (Figure 6.2).

6.2.2 ANTENNA DESIGN: FREQUENCY BANDS AND SIMULTANEOUS USAGE REQUIREMENTS

For smartphones, there are three frequency groups; low band (698–960 MHz), mid band (1710–2170 MHz), and high band (2300–2700 MHz). Antennas are usually designed to have three resonant structures: one for each frequency group. The bandwidth for the frequency groups are; 31.6%, 23.7%, and 16%, respectively. The antenna types used for smartphones typically have usable impedance and efficiency bandwidth of around 10%. This means for the frequency groups the antenna resonance is trying to cover more frequency bandwidth than the natural resonance of the structure will allow. Over the years, antenna designers have come up with many techniques to improve the bandwidth and efficiency of small antenna in handset implementation. In all cases, the antenna performance is a series of trade-offs to find the best performance compromise.

6.2.3 MULTIMODE, MULTIBAND PERFORMANCE

Instantaneous frequency requirements for the antenna structure in a smartphone are dependent on the geographical region where the phone is being used and the wireless service provider requirements in that region. In most cases, the operating frequency requirement calls for one primary band in one of the three frequency groups, and one other secondary frequency in a different frequency group. In CA, there may be primary bands in two frequency groups at the same time. In the future, there will be a requirement for two primary frequencies in the same group. Most antenna designs use triple resonant structures with fixed matching to cover all the bands of interest. With antenna aperture tuning, each resonance of the antenna structure can be independently tuned to a specific frequency band in the frequency group without affecting the performance of the antenna in another frequency group. This is a key capability of antenna aperture tuning that cannot be done with an impedance matching network. Antenna aperture tuning must be done on the part of the aperture that is resonant in the band of interest. Antennas can be designed where each part of the resonant structure operates independently, so tuning in one part of the structure does not affect the others.

6.3 ANTENNA VOLUME AND EFFICIENCY CONSIDERATIONS

6.3.1 THE SMALL ANTENNA LIMIT

The performance of small antennas is fundamentally limited and is a trade-off between size, bandwidth, and efficiency. This relationship was established by L.J. Chu based on models proposed by Harold Wheeler. Equation 6.1 shows the relationship of antenna volume (a^3) to wavelength (λ), efficiency (η), and fractional bandwidth ($\Delta f/f$).

$$\frac{\Delta f}{f} \propto \frac{(a/\lambda)^3}{\eta} \qquad (6.2)$$

This well-publicized relationship was developed in the 1940s has been shown empirically to reflect antenna performance for electrically small antennas.[1,2] This equation states that antenna size expressed in wavelengths has a direct impact on the antenna efficiency and bandwidth. If the antenna size is limited by industrial design constrains, then the instantaneous bandwidth and efficiency will also be limited.

6.3.2 BASELINE REQUIREMENTS FOR ANTENNAS IN SMARTPHONES

The general rule of thumb requirement for antenna performance in smartphones is 40%–50% radiation efficiency and maximum band edge voltage standing wave ratio (VSWR) of 3.5:1. These requirements, however, do not always translate in to acceptable total radiated power (TRP) and total isotropic sensitivity (TIS) performance. TRP and TIS are the actual measurements required by the service providers to allow the device to operate on their network. Increasing frequency range and industrial design constraints are making it difficult for a fixed-tuned antenna to cover the frequencies of interest with adequate performance.

6.4 MAXIMIZING EFFICIENCY: THE EFFECT OF LOADING AND UNLOADING THE ANTENNA

6.4.1 ANTENNA SYSTEM PARASITIC EFFECTS TO LOAD AND UNLOAD THE ANTENNA

In order to understand the effects of loading an antenna structure, it is important to understand how the antenna is integrated into the complete smartphone structure. Equivalent circuit analysis has been done and validated with simple blank printed circuit board (PCB) prototypes. Figure 6.3 illustrates a PIFA equivalent circuit model where the effect of the phone chassis is considered.[3]

This model shows the reactive elements of the antenna lumped into two terms, L_1 and C_1. In a real phone, the ground plane geometries and interactions are very complex. The antenna behaves like complex distributed network. Figure 6.4 shows how a PIFA structure is loaded by the ground plane nearby, where $\beta = \omega\sqrt{LC}$, $L_e = X\omega\sqrt{LC}$.

The ground nearby the PIFA creates capacitive loading on the antenna element. The dimensions of the radiating element affect the inductance and resistance of the distributed network. The distributed nature of this network means the impedance (voltage and current) is changing as the location moves along the length of the antenna. As a result, the impedance presented to the tuner varies along the length of the antenna. Choosing the right tap point to place the tuner depends on the capacitance range of the tuner, frequency range and radiation efficiency required to meet the performance goals.

For band select tuning, the fundamental frequency bandwidth must also be considered. This will establish the minimum dimensions of the antenna structure and is usually dominated by industrial design constraints. The fundamental frequency will

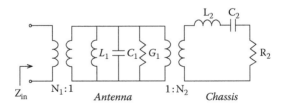

FIGURE 6.3   Antenna equivalent circuit model.

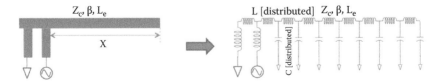

FIGURE 6.4   Loading of a PIFA structure and equivalent circuit.

also determine if the tuning element must present a capacitive or inductive reactance in order to tune the resonance of the antenna. If the fundamental resonant frequency of the antenna is at the low end of the frequency band, the reactance of the tuner must be inductive, to "pull" the frequency up. If the fundamental resonant frequency of the antenna is on the high end of the frequency band, the reactance of the tuner must be capacitive to "pull" the frequency down.

6.4.2 FREQUENCY SHIFTING AND TUNER PRECISION

As the tuner reactive load is moved farther out on the antenna structure, the impedance presented to the tuner increases. As the impedance increases, the loading effect of the reactance of the tuner increases, and it take less reactance to affect a resonant frequency change.

For capacitive loading, the fundamental resonance of the antenna structure is set to the high end of the frequency band. Increasing the capacitance will then lower the resonant frequency of the antenna. The capacitance range required to get the right frequency tuning range is determined by the location of the tap point on the antenna. If more tuning range is required, the tuner is placed farther out on the antenna.

6.4.3 EFFECT OF AMPLIFIER LOAD-PULL ON TUNING COMPONENT RESOLUTION

Power amplifier load-pull is a major consideration in final product acceptance testing. Small variations in the antenna tuning can have a big impact on the location of the impedance locus on the Smith chart. Figure 6.5 illustrates a typical amplifier load-pull diagram showing locations for optimum efficiency and output power.

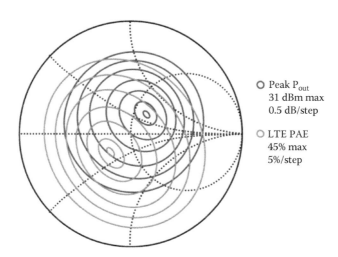

○ Peak P$_{out}$
31 dBm max
0.5 dB/step

○ LTE PAE
45% max
5%/step

FIGURE 6.5 Power amplifier load-pull diagram.

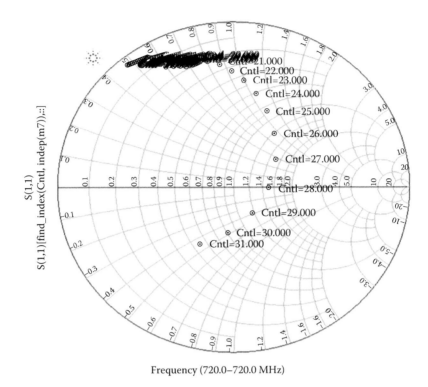

Frequency (720.0–720.0 MHz)

FIGURE 6.6   Locus of antenna impedance as the capacitive load shifts. (*Note*: Increasing control state corresponds to an increasing capacitance.)

Changing the resonance of the antenna causes the impedance locus on the Smith chart to rotate in a predictable way. With fine resolution in the tuning element, it is possible to make small favorable changes to the impedance not affecting the radiation efficiency but having a big impact power amplifier loading. The overall impact to power added efficiency of the RF chain can be several decibels (dB). The diagram in Figure 6.6 shows the trajectory of the antenna impedance as the resonant frequency is changed.

Moving the antenna impedance around the Smith chart allows it to pass through different regions of the power amplifier load-pull diagram changing output power and efficiency of the amplifier. Figure 6.7 shows the impact of impedance on TRP and power amplifier current, Idd.

6.5   TUNABLE COMPONENT PLACEMENT ON THE ANTENNA STRUCTURE

6.5.1   THE IMPACT OF CAPACITANCE RANGE ON RADIATION EFFICIENCY

In Section 6.4.1, it was shown that changing the value of the capacitive reactance along the antenna aperture changes the resonant frequency of a PIFA. It was also

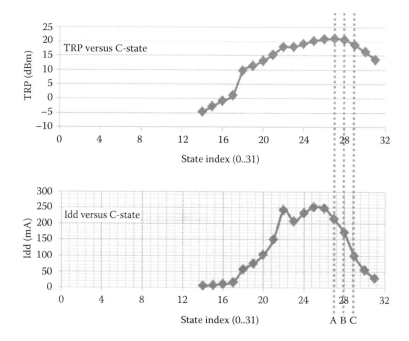

FIGURE 6.7 Total radiated power and power amplifier current over the capacitance range.

shown that the antenna aperture behaves as a distributed network. In principal, it seems there are a large number of choices for capacitance tuning range and device placement along the aperture to achieve a desired frequency tuning range. In practice, however, there are limitations on device placement that are driven by industrial design constraints and limitations on variable capacitor performance. To understand how these limitations affect device placement and capacitor tuning range, let us consider the voltage and current distribution along the length of the antenna. Figure 6.8 illustrates the voltage and current distribution on a typical PIFA structure.

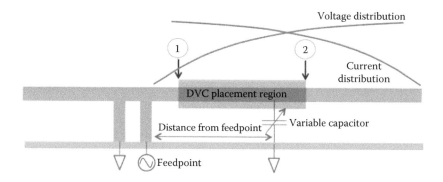

FIGURE 6.8 Voltage and current distribution along a typical PIFA structure.

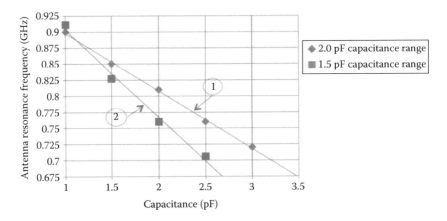

FIGURE 6.9 Tuning range as a function of the capacitance range. (*Note:* [1] is placed nearer the feedpoint; [2] is placed farther from feedpoint and as a result has a higher ΔFrequency/ΔCapacitance [tuning ratio].)

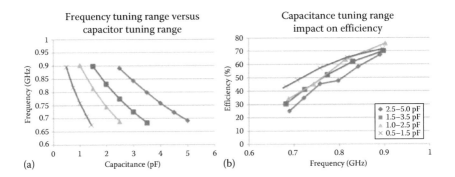

FIGURE 6.10 Tuning ratio: Impact of the capacitive range on efficiency and frequency tuning range. (a) Small capacitance ranges can produce an equivalent change in frequency compared to larger ranges if C_{min} is smaller. (b) Smaller capacitance loading on the antenna produces higher radiated efficiency while still covering the desired frequency range.

From the illustration in Figure 6.8, moving farther out on the PIFA increases the capacitive loading effect of the tuner and increases the voltage at the terminals of the tuner. At increased voltage, the loading of the capacitor will have a larger effect on the resonant frequency; therefore, it is possible to get the same frequency tuning range for different capacitance tuning ranges depending on where the variable capacitor is placed along the length of the antenna. Figure 6.9 shows two different capacitor ranges at different location on the antenna producing the same frequency tuning range.

Although it is possible to get the same frequency tuning range using different capacitor values, the maximum radiation efficiency is achieved when using the smallest capacitive loading values. Figure 6.10a shows the same frequency tuning range achieved with different capacitor ranges and the impact on radiation efficiency (Figure 6.10b).

6.5.2 The Impact of Quality Factor (Q) on Radiation Efficiency

The Quality factor (Q) of the tuner is a critical performance parameter. Refer back to Equation 6.1 for the relationship between R, L, and C in determining Q. Q is inversely proportional to the device ESR and therefore the implementation loss of the device. As the Q of the device drops, the ESR increases for a given capacitance value and, the implementation loss increases. The chart in Figure 6.11 illustrates how ESR affects the radiation efficiency of the antenna.

MEMS technology makes it possible to build tuners with high $Q > 100$ and low ESR <0.5 Ohm, thereby reducing the implementation loss to fractions of a dB.

6.5.3 The Impact of C_{min} on Radiation Efficiency

The minimum capacitance presented by the tuner is a critical performance parameter. The lower the C_{min}, the higher the efficiency will be. Let us consider the distributed circuit model of a PIFA introduced in Section 6.4.1. The addition of shunt capacitance lowers the resonant frequency of the structure. When a tunable capacitor is added to an antenna structure, even a small amount of capacitance will lower the resonant frequency. To compensate, the antenna length must be shortened. This reduces the antenna size and lowers the efficiency. The ideal tunable capacitor will have a zero minimum capacitance so that the unloaded antenna length and efficiency can be maintained. In addition to ESR, implementation loss is also a function of capacitance. As capacitance increases, the effect of ESR also increases. Depending on where the variable capacitor is attached, it is possible to get a wide frequency tuning range with almost any capacitance range. The selection of the right capacitance range is a function of industrial design (where the tuner can be placed) as well as radiation efficiency. It is always desirable to keep the capacitance loading to a minimum to keep the radiation efficiency as high as possible.

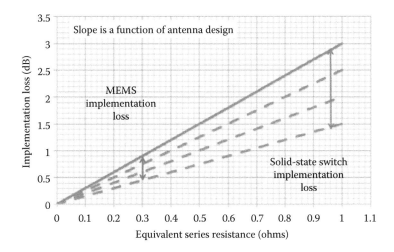

FIGURE 6.11 Impact on radiation efficiency as a function of ESR.

6.5.4 LINEARITY CONSIDERATION IN TUNER PLACEMENT

Moving the tuner out on the antenna makes it possible to use the smallest amount of capacitance to affect tuning range. This has the benefit of keeping the radiation efficiency as high as possible. However, the voltage presented at the terminals of the tuner can be quite high. High voltage on the tuner tends to increase the second- and third-order harmonics generated by the tuner. MEMS device construction makes it possible for the tuner to withstand high voltage (>40 V RMS) without generating unacceptably high harmonics.

6.5.5 VOLTAGE AND IMPEDANCE CONSIDERATIONS

The placement of the tuning element on the antenna aperture impacts the performance. Moving farther out on the antenna element puts the tuning device in a high impedance region of the antenna. At higher impedance, a small change in capacitance will cause a big change in resonant frequency. This is good since it allows the antenna to remain as large as possible. However, as the device is placed farther from the feedpoint and the impedance increases, the voltage across the device will increase. At wideband code division multiple access (WCDMA) and LTE power levels of +24 dBm, the RMS voltage at the terminals of the tuner is generally below 30 VRMS even when the tuner is placed at 15 mm or farther from the feedpoint. At higher power levels, like GSM power of +33 dBm, the voltage across the terminals of the antenna can approach 70 VRMS at a distance of 15 mm from the feedpoint. At these voltage levels, many solid-state devices have poor linearity and exhibit high second- and third-order harmonics at voltage levels as low as 20 VRMS. MEMS devices are more tolerant to higher voltages and can be used in regions of the antenna where voltage is as high as 40 or 50 VRMS. This allows the tuner to be placed in a region where smaller capacitance values can be used to tune the antenna, thus preserving as much of the antenna length and efficiency as possible.

The graph in Figure 6.12 shows the RMS voltage at the terminals of the tuning device for various positions along the length of the antenna. The dashed traces are at 0.4 pF

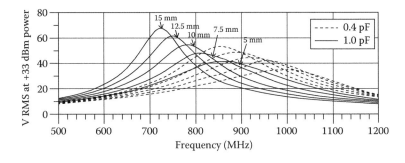

FIGURE 6.12 Antenna voltage at 33 dBm with different tuner locations. (*Note*: At 15 mm from the feedpoint, nearly 70 VRMS is generated on the tuner.)

and the solid traces are at 1.0 pF. The rightmost trace of each group is at 5 mm from the feedpoint and the leftmost trace is 15 mm. The increment between traces is 2.5 mm.

6.6   SMARTPHONE TUNABLE ANTENNA DESIGN

6.6.1   Antenna Design Considerations in Modern Smartphones

In the past, it was common for a cell phone antenna to be a separate part made from stamped metal, flexible printed circuit material, or laser direct structuring onto a plastic housing. The antenna was then integrated into the phone as part of the assembly process. Today, smartphone designs are dominated by large touch screens, forward and rear facing cameras, large batteries, and thin form factors. In many cases, there is no room for a separate antenna component. In addition to the lack of space, metal frames are often used to make smartphones mechanically rigid to avoid breaking the display. This combination of factors makes it difficult to accommodate traditional antenna designs. Recently, antenna designers have started to use the structural elements of the phone, metal frames and metal backs, as the antenna radiating element. In fact, the advent of antenna aperture tuning devices makes it possible to "tap" into an existing mechanical structure and drives it as a radiator.

6.6.2   Metal Frame Antenna Design

The example in this section identifies the design considerations for a metal frame antenna implemented in a typical smartphone form factor. Figure 6.13 shows the geometry of the phone and metal frame.

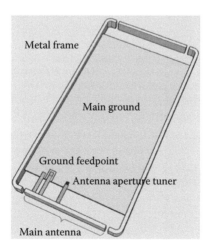

FIGURE 6.13   Typical metal frame smartphone.

TABLE 6.1

Comparison of Digital Variable Capacitors (DVC) Used for Design

			Parameters at the Board Side of Bump Pads			
			Capacitor Values Used—of 32 Possible States (pF)			
Device	R (Ohm)	L (nH)	C_{min}	C10	C20	C_{max}
32CK301	0.3	0.8	0.45	0.64	0.84	1.05
32CK417	0.4	0.8	0.55	0.92	1.29	1.70
32CK402	0.4	0.8	0.70	1.14	1.57	2.05

The frame is 65 mm wide × 130 mm long × 8 mm thick. There is a 10-mm clearance between the main ground plane and the frame on either end of the phone. The frame has four breaks to create four distinct sections. The frame sections on the side of the phone are connected to the main ground plane. The two frame sections on either end are floating. The device in Figure 6.13 uses the bottom frame edge as the main antenna radiator. This piece of the frame has three connection points, ground, feedpoint, and antenna aperture tuner to form a tunable PIFA structure.

Several available 32-states MEMS aperture tuning devices were selected to perform the study. A list of the devices and performance parameters are listed in Table 6.1.

Each of the three MEMS devices was connected to the main antenna radiator in a position to tune the antenna between 700 and 800 MHz. The data shows that each of the devices can be used to tune the antenna over the same range of frequencies by tapping into the antenna at the right point. The 32CK301 device has the lowest C_{min} and smallest capacitance tuning range. From the information in Section 6.5.1, it stands to reason that this device will be placed the farthest from the feedpoint in order to get a sufficient tuning range. The proper device distance is around 15 mm. The 417 has slightly higher C_{min} and wider tuning range, it is connected to the main antenna 10 mm from the feedpoint. The 402 has the highest C_{min} and widest tuning range and is connected to the main antenna 5 mm from the feedpoint. S11 Log Mag and efficiency were measured for each device at four different capacitance states, C_{min}, C10, C20, and C_{max} (Figure 6.14).

The C_{min} of the capacitor will cause a shift in the frequency response of the antenna. This is the effect of adding an aperture tuning device. A device with a low C_{min} will minimize this effect. From the data, each of the three MEMS variable capacitors has a similar shift in frequency even though they have different C_{min} values. The reason is that each of the device, in order to compensate for a different capacitance tuning range is placed at a different position on the antenna. The device with the smallest capacitance tuning range is placed farther out on the antenna where the impedance of the antenna is higher. In this region, even a small capacitance will have an impact on the resonant frequency. Devices withs larger tuning range can be placed closer to the feedpoint and achieve the same frequency tuning range, but in the case of the selected MEMS devices, each has a different C_{min} value as well. The change in C_{min} and location together tend to balance out giving roughly the same frequency shift for each device (Figure 6.15).

The impact of ESR was studied using a combination of discrete capacitors and resistors. The capacitor is placed in parallel across the aperture of the antenna, the same way the MEMS device is placed on the antenna. A discrete resistor is then placed in series with the capacitor to simulate a device with a higher ESR. The objective of the study is to show the impact of ESR on radiation efficiency.

Figure 6.16 shows a mockup of the frame antenna with a discrete capacitor and series resistor in place of the MEMS aperture-tuning device.

The resistance is changed from 0 to 2 ohms. In each case, the return loss (S11) magnitude and efficiency are measured. The experiment was performed using a 2.7 pF capacitor at a distance of 5 mm from the feedpoint.

Figure 6.17 shows the S11 measured results and Figure 6.18 shows the measured efficiency.

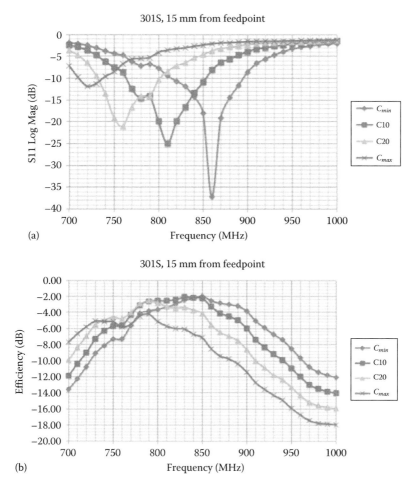

FIGURE 6.14 Comparison of return loss (S11) (a) and efficiency (b) for different DVC devices and locations. (*Continued*)

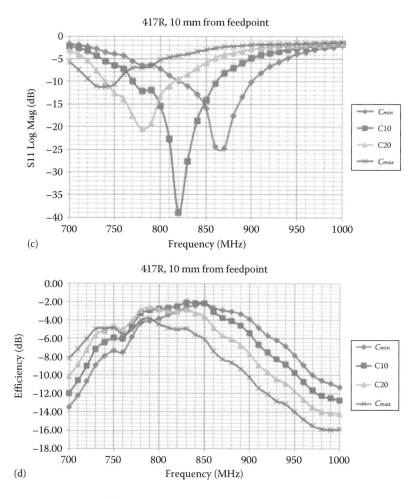

FIGURE 6.14 (*Continued*)   Comparison of return loss (S11) (c) and efficiency (d) for differ-ent DVC devices and locations.   (*Continued*)

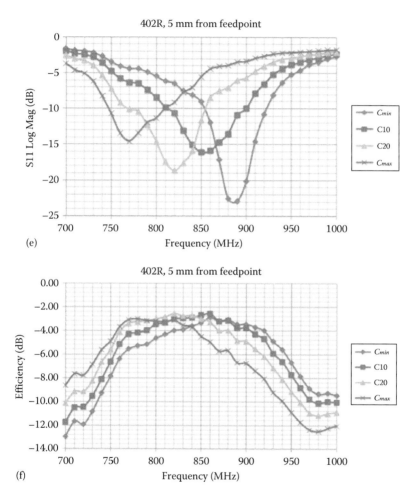

FIGURE 6.14 (*Continued*) Comparison of return loss (S11) (e) and efficiency (f) for different DVC devices and locations.

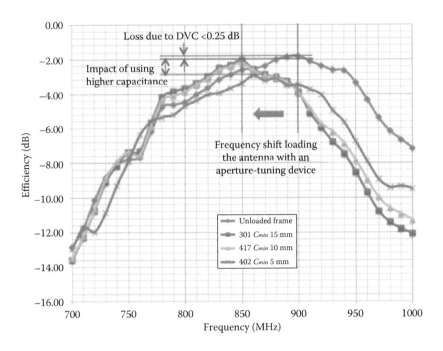

FIGURE 6.15   Analysis of capacitance loading of the tuning devices.

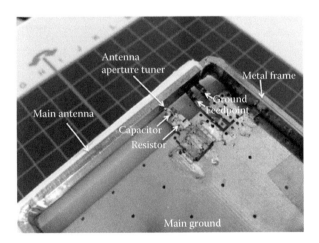

FIGURE 6.16   Development platform for measuring return loss and efficiency.

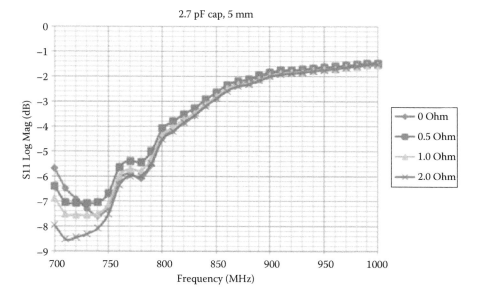

FIGURE 6.17 Return loss (S11) magnitude shown on a logarithmic scale.

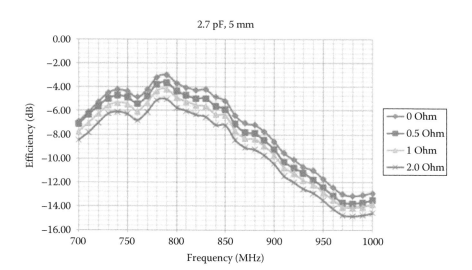

FIGURE 6.18 Measured radiated efficiency.

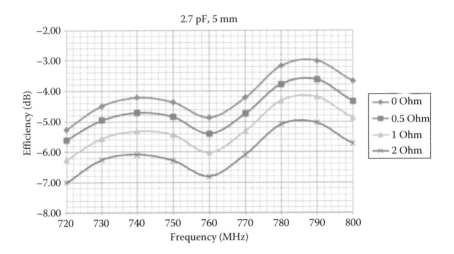

FIGURE 6.19 Magnified scale showing antenna efficiency impact of ESR over frequency.

The chart in Figure 6.19 shows the measured efficiency with a magnified scale in order to more clearly see the impact of high capacitance and high ESR on the radiation efficiency.

6.7 CONCLUSIONS

Antenna design challenges in smartphones are driven by industrial design constraints and a proliferation of frequency bands of operation. Designing aperture-tuned antennas to meet the challenge offers significant advantages in total radiated power and receiver sensitivity over fixed frequency and variable impedance-tuned antenna designs. When designing an aperture-tuned antenna, it is important to understand the unique design requirements and select the appropriate tuning element.

The performance characteristics of RF MEMS are particularly well suited to aperture-tuned antennas. RF MEMS devices have very low equivalent series resistance (ESR). The typical RF MEMS capacitor has an ESR from 0.2 to 0.5 ohm. This parameter is particularly important as it is directly related implementation loss. As the ESR increases, the implementation loss increases. ESR > 0.5 ohm can lead to implementation losses from 1 to 3 dB. This much loss is unacceptable in the front end of the RF chain. In addition to ESR, minimum capacitance or C_{min} is also a critical parameter. The fundamental antenna structure acts as a distributed network of R's, L's, and C's. A shunt variable capacitance placed in the antenna aperture will load the radiating element and lower the resonant frequency. To compensate, the antenna must be made shorter. To preserve antenna length and maximum radiation efficiency, the antenna must be loaded as little as possible. A very low C_{min}, <0.5 pF, allows the fundamental antenna to be as big as possible preserving the radiation efficiency.

REFERENCES

1. Wheeler, H.A., Fundamental limitations of small antennas, *Proceedings of the IRE*, 35, 1497–1484, December 1947.
2. Chu, L.J., Physical limitations of omnidirectional antennas, *Journal of Applied Physics*, 19, 1163–1175, December 1948.
3. Vainikainen, P. et al., Resonator-based analysis of the combination of mobile handset antenna and chassis, *IEEE Transactions on Antenna and Propagation*, 50(10), 1433–1444, October 2002.

7 RF-CMOS Impedance Tuners

Performance Metrics and Design Trade-Offs

Tero Ranta

CONTENTS

7.1 INTRODUCTION—TOWARD A FULLY INTEGRATED RECONFIGURABLE CELLULAR RF FRONT END

The complexity of cellular devices has increased at a very rapid rate due to the exponential growth in mobile data demand. The use of more spectrally efficient higher-order modulation schemes has enabled transmitting and receiving significantly more bits per Hertz of radio spectrum, but additional frequency bands must be added to the cellular device to cover global operation and various frequency band allocations. This spectrum growth not only increases the number of duplex filters, antenna switch throws, and antennas, but also requires fundamental modifications to the radio frequency (RF) front-end architecture to support flexibility and reconfigurability.

Monolithic integration using complementary metal oxide semiconductor (CMOS) technology has been a growing trend in the baseband, application processor, and transceiver areas of the cellular device. Over the past several years the CMOS revolution has also swept the cellular RF front end, enabled by significant technological advancements in silicon-on-sapphire (SOS) and silicon-on-insulator (SOI) technologies. The growing use of CMOS SOI technology in the RF front-end components, such as antenna switches and power amplifiers also makes it a natural candidate for implementation of impedance tuning devices. This enables monolithic solid-state integration for a truly fully integrated reconfigurable CMOS SOI solution and provides a small die area, good cost structure, high yield, and high reliability required for very high-volume mass production.

Key cellular applications for tuning devices include antenna impedance tuning[5-7,10]; antenna aperture tuning[8]; power amplifier output impedance tuning; tunable phase shifters[9]; duplex filter tuning; and diplexer, coupler, and low-pass filter tuning. Typical specification requirements for tuning devices used in these applications can be summarized,[2-11] as supporting capacitance ranges within 0.5–30 pF, tuning ratios from 3:1 to 30:1, Quality factors in the range of 20–80 at 1–2 GHz, ability to withstand instantaneous peak RF voltages of 15–90 V_{PK}, and RF power levels above +35 dBm, being able to reliably handle 10–100 billion switching cycles and being controlled using a high-speed serial digital interface such as MIPI RF front-end control interface (RFFE).

7.2   FUNDAMENTALS AND DESIGN TRADE-OFFS OF SEMICONDUCTOR SWITCHED CAPACITORS

There are several ways to implement cellular RF impedance tuning on a CMOS process, including stacked field effect transistor (FET) switched capacitors, stacked varactor diodes, and microelectromechanical system (MEMS) actuators. The main focus of this chapter is on solid-state switched capacitor implementations, given their significant ruggedness and reliability, wide tuning ratio, integration potential, and ease of use. Even though on a conceptual level, switched capacitors are simple devices, significant innovation has gone into taking them from a concept to commercially viable and cost-competitive devices that meet the stringent requirements of modern cellular devices.

7.2.1   BASIC CONCEPTS AND FUNDAMENTAL TRADE-OFFS

The basic concept of digitally tunable capacitor (DTC) devices implemented on RF-CMOS SOI process is the use of high-linearity stacked FET RF switches to switch on-chip capacitors in and out of the circuit, via a digital control interface (Figure 7.1a). Since the sizes of these switches and capacitors are usually binary weighted and constant Q, the switched capacitor network can be thought of as a very linear digital-to-capacitance converter. The FETs are turned ON and OFF by applying a positive or negative voltage via a digital control interface, respectively, on their gate terminals through series biasing resistor.

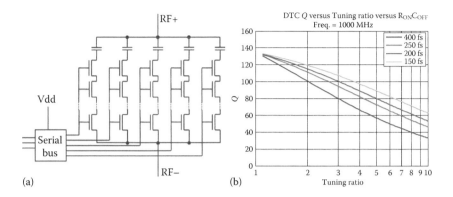

FIGURE 7.1   Basic switched capacitor implementation (a) and tuning ratio versus Quality factor trade-off versus $R_{ON}C_{OFF}$ (b).

All solid-state switched capacitor implementations are governed by the same fundamental trade-offs between the capacitance tuning ratio and Quality factor (Figure 7.1b), which stems from using switch transistors in series with fixed capacitors. The main figure of merit for these switches is the $R_{ON}C_{OFF}$, which refers to the product of ON-state resistance and OFF-state capacitance (in units of femtoseconds), which is ideally independent of the FET dimensions or number of FETs in a stack. The pace of RF-CMOS SOI development has enabled significant year-over-year reductions in $R_{ON}C_{OFF}$ values, akin to the ubiquitous Moore's law for CMOS digital circuits. The first commercial implementations of DTC technology on SOS substrate used FET devices with $R_{ON}C_{OFF}$ values above 700 fs, whereas modern RF front-end SOI CMOS processes exhibit $R_{ON}C_{OFF}$ figure of merits close to 100 fs.

The fundamental capacitance, Quality factor and S_{11} behavior is illustrated in Figure 7.2, for a 5-bit 0.9–4.6 pF 30 V_{PK} DTC implemented on an SOS process. The Quality factor curves shapes illustrate almost "ideal" fundamentally stacked switched capacitor behavior. In maximum capacitance state (C_{31}) the behavior is dominated by series resistance of ON FETs (R_{ON}) in series with the metal–insulator–metal (MIM) capacitor (C_{MIM}) resulting in the familiar inverse of frequency (1/f) slope for Q. In minimum capacitance state (C_0), all FETs are OFF (and appear as capacitors C_{OFF}) and the high-frequency Q slopes are dominated by the series connection of MIM capacitor Q and that of FET's OFF capacitance (Q of C_{OFF}). The frequency where C_0 and C_{31} curves intersect is known as the Q crossover frequency, and typically targeted below the minimum typical operating frequency. All of the Q slopes tend to 0 as frequency goes to 0. This is because the biasing resistors in the FET stack become effectively parallel with total capacitance of the DTC, resulting reduction in Q (also known as *de-Q'ing*). Additionally, the substrate loss in SOI processes typically limits the Q performance at low-capacitance states, causing the Q low-frequency slope to flatten out significantly and impact the Quality factor at 1–2 GHz. SOS processes do not have this behavior, due to the extremely low-loss tangent substrate material. The Quality factor for states

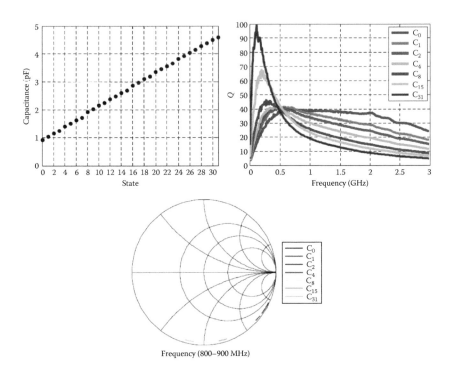

FIGURE 7.2   Fundamental C, Q, and S_{11} behavior for a 0.9–4.6 pF digitally tunable capacitor.

between C_0 and C_{31} is a combination of the two behaviors depending on how many bits are ON and OFF.

The main differentiator between RFSOI and bulk CMOS switched capacitor implementations is that SOI enables the use of FET stacking to arbitrarily increase the RF peak voltage handling (V_{PK}) of the tunable component to levels required by cellular devices. This is made possible by the use of a fully insulating SOI (or SOS) substrate, which enables nearby FETs to float with respect to each other, enabling increase of voltage handling from 2 to 4 V_{PK} for a single FET to up to 100 V_{PK} for the whole stack. This allows the RF power handling capability to be fully tailored to the specific application requirements by simple circuit design techniques, rather than having to modify materials properties of dielectrics or mechanical properties of moving parts.

The overall die area of the switch FET stack is proportional to the square of stack height in order to maintain the same total DC resistance across the stack regardless of the stack height (i.e., going from A stack of one to two requires doubling the size of each device to maintain constant total resistance, which quadruples the total die area). This causes the die to grow very rapidly with increase in V_{PK} specification. In practice this is limited by (1) parasitic capacitance resulting from large die area and (2) limited available die area, which bounds the achievable C_{min}, V_{PK}, and FET stack DC resistance (i.e., Q).

7.2.2 REAL-LIFE PERFORMANCE IS DOMINATED BY PARASITIC EFFECTS

The industry drive toward lower $R_{ON}C_{OFF}$ figures-of-merit (by optimizing RFSOI process and device parameters) is motivated by improving loss and bandwidth performance for RF switch applications. The downside of this trend is a continued reduction in a single FET's breakdown voltage, or ability to hold off RF voltage in the OFF state, which is detrimental to tuning devices that usually require much higher peak voltage (V_{PK}) specifications compared to typical RF switches. Lower per-FET breakdown voltage requires increasing the stack height to maintain the same total peak voltage (V_{PK}) specification, which also requires making each FET larger to reduce its resistance to maintain the same total R_{ON} for the stack. The increase in stack height and device sizes significantly increases the impact of unwanted parasitics and degrades the effective stack-level $R_{ON}C_{OFF}$. Careful process and design optimization is required to find the optimal FET device parameters for tuning applications.

The lowest achievable minimum capacitance (C_{min}) and hence the capacitance tuning ratio are dominated by the OFF capacitance of the FET stack relative to the MIM capacitor value. Ideally, the total stack-level C_{OFF} would be independent of the stack height, but in reality each FET in the stack has a parasitic capacitance (C_P) to ground which is typically 1%–2% of C_{OFF}. This distributed parasitic capacitance limits the minimum achievable stack-level C_{OFF} regardless of how many FETs are connected in series. The product level C_{min} is further increased by parasitics due to packaging and die interconnect.

The Q factor of the switched capacitor circuit is dominated by R_{ON} of the FET stack, which includes the DC resistance of each FET and distributed metal routing RF resistance along the length of the FET stack. The Q factor is further reduced by the MIM capacitor dielectric and electrode losses (typically $Q = 100$ @ 2 GHz) and metal interconnect resistance (typically 0.1–0.2 ohms). The equivalent SRF (self-resonant frequency) is dominated by the distributed inductance of the FET stacks, metal routing, and bond wires or flip-chip bumps.

To capture these parasitic effects, a three-terminal product level equivalent circuit of the DTC can be created (Figure 7.3), where the element values depend on the tuning state. The total parasitic capacitance to GND (C_{P1}, C_{P2}) is a significant portion of the series capacitance (C_S) and must be taken into account when designing tuning networks using DTCs in series configuration.

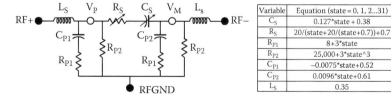

Variable	Equation (state = 0, 1, 2...31)	Unit
C_S	0.127*state + 0.38	pF
R_S	20/(state+20/(state+0.7))+0.7	Ω
R_{P1}	8+3*state	Ω
R_{P2}	25,000+3*state^3	Ω
C_{P1}	−0.0075*state+0.52	pF
C_{P2}	0.0096*state+0.61	pF
L_S	0.35	nH

FIGURE 7.3 Example of a product level equivalent circuit model including all parasitic effects.

Another major effect of the distributed parasitic shunt capacitance is that it causes the RF voltage to divide unevenly across the FET stack. The overall voltage handling of the switch FET stack is limited by the weakest device (highest relative RF voltage in the stack). Due to the nature of a series-shunt capacitance ladder, the FETs on top of the stack see the highest relative voltage across them and start clipping the RF signal first resulting in high distortion. The voltage division must be equalized in order to get the full peak voltage handling performance, especially for stack heights above approximately 8 as uneven voltage division increases almost exponentially with increasing stack height. Often tunable devices need to support both shunt and series operation. This makes it even more challenging to achieve high-peak voltage handling, as the FET stack may experience highest per-device voltages on top or bottom of the stack, depending on whether the terminals see high- or low-impedance levels when being used in the tuning network. It is very important to codesign the tuning network and tunable capacitors as a whole to optimize performance of the whole network.

Another side-effect of increasing stack height is that it makes the circuit more susceptible to unintended Q degradation (*de-Q'ing*) due to the bias resistors connected to the gates of the stacked FETs. The Q degradation can be reduced by increasing value of the gate resistor at the expense of switching time which is dominated by the RC time constant between gate bias resistor and FET gate capacitance. To further reduce the de-Q'ing effect, the MIM capacitors are usually placed on top of the FET stack for shunt connected devices making them "invisible" to the gate resistors. MIM capacitors may be placed at both ends of the FET stack if full DC blocking is required in series configuration, but this typically requires adding a separate DC path to the FETs between the series capacitors, which will cause additional de-Q'ing.

The main benefit of using SOS substrate is that it is perfectly insulating with a very high Quality factor. Typical HR SOI (high-resistivity SOI) substrates exhibit higher parasitic capacitance with lower Quality factors for the parasitic substrate network ($Q = 15$–30), which will impact especially high V_{PK} and high-frequency implementations on SOI.

The CMOS FET devices and MIM capacitors used in switched capacitors are sensitive to electrostatic discharge (ESD) conditions with typical ratings in several kV of human body model (HBM). This is much better than a typical GaAs switch device with some hundreds of volts of HBM tolerance. However, the typical cellular device specification for any device connected directly to the antenna port is 8 kV HBM and there are additional surface-mount device (SMD) assembly process related machine model (MM) and charged device model (CDM) requirements for standalone components. Guaranteeing this level of ESD robustness not only requires providing high-linearity low-capacitance ESD discharge paths on-chip between all external terminals of the tunable device, but also making sure the instantaneous voltage across any on-die MIM capacitor does not exceed the breakdown voltage (~50–100 V) of the dielectric. The switch FET stack is self-protecting due to the drain-source breakdown of the FETs during ESD event, but ESD damage on MIM capacitors usually results in short circuits in the dielectric near weak spots in the dielectric layer. Fully integrated tunable matching networks have typically less stringent ESD requirements due to less exposed RF ports and integrated shunt inductors on RF ports.

7.2.3 COMPONENT QUALITY FACTOR

One of the important figures of merit of tunable components is the Quality factor, as it directly impacts the matching loss and ability to match loads with high-reflection coefficients or high-resonator Quality factors. The fundamental definition is based on energy stored versus energy lost per cycle of the RF waveform. The more commonly used definition for series Quality factor is $Q = \Im(Z)/\Re(Z)$, that is, the ratio of imaginary part of the shunt input impedance (reactance) to the real part (resistance). However, this definition cannot be used for resonant circuits as at resonance the positive inductive reactance would cancel out the negative capacitive reactance and result in an incorrect Quality factor of zero.

Resonant circuits (i.e., circuits containing resistance [R], inductance [L], and capacitance [C]) require the use of a resonator Q method, as illustrated in Equation 7.1. The resonator Quality factors for series and parallel RLC circuits are as follows:

$$Q_{ser} = \frac{1}{R}\sqrt{\frac{L}{C}} = \frac{\omega L}{R} = \frac{1}{\omega RC} \qquad Q_{par} = R\sqrt{\frac{C}{L}} = \frac{R}{\omega L} = \omega RC \qquad (7.1)$$

The resonator Q method requires decomposing the total reactance to capacitive (C) and inductive (L) components, and separately calculating the total Quality factor using one or the other (same result independent of which reactance is used). Determining this based on a measurement requires input impedance measurements at least at two different frequencies. An elegant method to determine resonator Q is to measure phase of input impedance (Z_{IN}) at two different frequencies around the resonant frequency and calculate rate of change on phase of input impedance as a function of normalized frequency as shown in Equation 7.2 (phase measured in radians). The resonator Quality factor from the measured input impedance phase is as follows:

$$Q_{res} = f_{res}\frac{\Delta\varphi_{Z_{IN}}}{\Delta f} = f_{res}\frac{\left|\varphi(Z_{f2}) - \varphi(Z_{f1})\right|}{\left|f_2 - f_1\right|} \qquad (7.2)$$

An alternative method is to calculate C from Z_{IN} measured at very low frequency (almost DC) as it is not impacted by series inductance yet. Then use the "DC" value of the capacitance along with resistance measured at the *RF* to calculate the Q. This is identical to eliminating the series L completely and only using the true capacitance to calculate Q.

For measuring the Quality factor of tunable components, an RF impedance analyzer or *LCR* meter that uses the RF-IV technique and supports measurements up to 3 GHz should be used. This method is based on measuring the RF voltages and currents into the load, which is ideal for measuring highly reflective loads near the edge of the Smith chart (such as high-Q capacitors and inductors in one-port configuration). In contrast, the vector network analyzer (VNA) is a reflectometer-based instrument using swept sinewave RF sources and directional couplers to measure forward and reflected RF waves. This instrument is optimized for measuring impedances in the vicinity of 50 Ω and will have a lot lower measurement accuracy compared with an RF impedance analyzer. When measuring tunable capacitors, great care must be taken to properly de-embed the coaxial connectors and RF transmission line leading up to the device all the way up to the proper reference plane (i.e., flip-chip bumps or other interconnect). Any small difference (on the order of picoseconds) in electrical delay can impact the phase relationship between the measured real and imaginary part of the input impedance, thereby having a large impact on capacitance and Quality factor responses.

7.3 IMPEDANCE TUNING NETWORKS

7.3.1 Tuning Network Circuit Topology and Trade-Offs

One of the most important specifications for a tunable matching network is to transform a range of load impedances (e.g., VSWR = 4:1) to a range of input impedances (e.g., VSWR = 1.5:1) across the required frequency band (e.g., 700–900 MHz), while incurring as little matching loss as possible. Classical matching network design theory is based on calculating the values of lossless fixed components assuming static port impedances. Tunable matching networks must cover wide frequency and impedance ranges using low-Q components, requiring numerical optimization of component values and tuning ranges to meet specifications. Once the matching network topology (low, high, or bandpass) has been selected, additional pass-band shaping (harmonic notch filters, etc.) can be added with appropriate component value changes to account for loading effects. A typical 3D bandpass tuning network with component losses included is illustrated in Figure 7.4.

Adding more sections (degrees of freedom) to a tunable matching network increases its impedance coverage in theory, but, in practice, this is severely limited by increased loss of the added sections, which reduces impedance coverage. Thus the number of sections needs to be carefully tailored to the application. Given the fundamental trade-off between the solid-state switched capacitor tuning ratio and Quality factor, having a tuning network simultaneously cover a wide load impedance and frequency range also incurs higher dissipative loss than an equivalent circuit with less frequency or tuning range.

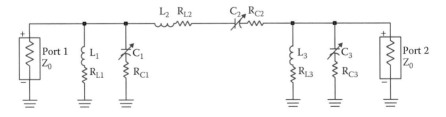

FIGURE 7.4   3D impedance tuner with lossy components.

While placing tunable elements near the high impedance areas of the tuning network or antenna allow for wider impedance coverage, the downside is that the RF peak voltage (V_{pk}) across the tuning element increases significantly. Switched capacitor networks usually have a very strict limit on the maximum RF voltage. On the other hand, placing the tunable element near low impedance area reduces the peak voltage but, at the same time, increases the current element resulting in higher power dissipation. Careful consideration is required to place the tunable components in a location that maximizes their performance.

Additionally, tunable capacitors used in series configuration often experience high RF voltage levels, due to voltage multiplication and resonant conditions. Using elements in shunt configuration only is significantly easier but reduces frequency and impedance coverage of the tuning network.

From a network circuit topology point of view, the lower the source or load port impedance, the higher the current through the circuit elements. The power dissipation is proportional to the square of the current and, thus, increases rapidly for low port impedances. This is especially critical for impedance transforming matching networks often used to match amplifier devices to 50 Ω system impedance.

Regardless of the intrinsic Quality factor of the tunable component, the network loss is determined by the mounted Q factor when the component is used in the actual tuning circuit. Due to parasitic inductance, capacitance, and resistance, the network Q is always much lower than the tunable element component Q at GHz frequencies, as parasitic impedances have values comparable to the actual impedances. Even more important is the actual total dissipative loss of the tunable matching network when it is being used to match impedances. The overall loss is composed of several different factors. Dissipated power loss of an individual component can be calculated as half of the real part of the conjugate of complex voltage drop multiplied by the complex current (Equation 7.3). The power dissipation for two ports is the difference between power entering port 1 and power leaving port 2 (Equation 7.4).

Component real power dissipation:

$$P_{comp} = \frac{1}{2}\Re((V_2 - V_1)^* I) \tag{7.3}$$

Network real power dissipation:

$$P_{network} = \frac{1}{2}\Re(V_1^* I_1) - \frac{1}{2}\Re(V_2^* I_2) \tag{7.4}$$

A Pareto chart can then be created showing the contribution of each element to the total loss of the network. This usually highlights the importance of reducing substrate loss effects, impact of finite inductor Q, and importance of minimizing EM (electromagnetic) effects, such as current crowding, skin effect, spreading resistance, proximity effect, and radiation loss.

The best measurement tool for impedance tuning network S-parameters is the VNA, even though some custom automation software usually needs to be written to enable collection of hundreds of thousands or millions of states.

For large signal characterization of parameters such as harmonic and intermodulation distortion (IMD), as well as peak voltage handling, the impedances and capacitive and inductive parasitics of the tuning network and individual components must be well known. Distortion produced by the tuning network cannot be easily predicted from tunable capacitor harmonic and IMD specifications, because nonlinear currents and voltages generated by tunable capacitors are significantly impacted by the rest of the tuning network. The effective peak RF voltage seen by the switch FET devices must not exceed their design and reliability limits under typical operation of the tuning network. Depending on configuration (shunt, series) and port impedance levels and any circuit resonances, the voltage across the FET stack may be several times higher than effective RF voltage at the ports calculated from the available RF power levels.

7.3.2   CELLULAR SYSTEM LEVEL CONSIDERATIONS

There are several parameters that require careful consideration when implementing tunable impedance matching networks for modern cellular devices. Modern wideband communication systems with high-order modulations (e.g., QAM, OFDM, LTE, and 802.11ac) use lower average power (typically <+24 dBm at cellular device antenna port) compared with 2G modulation standards (e.g., up to +35 dBm for GSM). For high peak-to-average waveforms, the high voltages at the peaks must be taken into account when determining the peak RF voltage rating for the tuning network. Another system level concern are the rapid transmission phase changes during tuner state changes, which will require allowing only relatively small changes timed to occur during the safe periods of the full-duplex FDD waveforms (e.g., frame/slot boundaries). For closed-loop systems, the control algorithms must be made robust and adaptive against oscillation between adjacent states. This typically requires the use of gradient-free discrete optimization algorithms. Finally, the linearity requirements are extremely stringent for carrier aggregation (CA) systems to support a multitude of simultaneous carriers and band combinations.

7.3.3   DEALING WITH EXPONENTIAL COMPLEXITY

Monolithic integration enables adding a lot of tunability without necessarily incurring any significant cost. The downside is the more reconfigurability and degrees of freedom are added, the more characterization and test is required to cover all the unique tuning states.

This is best illustrated by way of an example. Tuning network with four tunable capacitor at 5-bit resolution each (i.e., 4D 5-bit) will have a total of over 1 million (2^{20}) unique tuning states. Assume a two-port VNA sweep at 300 frequency points is required for characterization purposes. Every two-port S-parameter frequency point requires storing of four complex floating point numbers for S_{11}, S_{12}, S_{21}, and S_{22} in the form of eight real and imaginary components. The total number of complex S-parameter points becomes 1.25 billion ($300*8*2^{20}$)! Even at a typical VNA sweep times of 100 ms (including data transfer and storage overhead), it would still take about 30 h to sweep across the whole tuning space. The number of measurements could be significantly reduced by, for example, measuring every second or fourth tuning state, however, narrow resonances may be missed completely.

Regular S-parameter files are stored in ASCII format, which for two-port files amounts to about 128 bytes of data for every frequency point (16 bytes per floating point number). The total file size would be over 40 GB for 1 million tuning states and would take several hours to load into a circuit simulator, which is prohibitively inefficient. A good alternative is to store the file in a binary format. Not only does this reduce the file loading time to a few seconds, there is also no waste because data are stored natively in the most efficient way. A typical double-precision floating-point format (IEEE754) uses 8 bytes (64 bits) per number, cutting the file size in half to 20 GB. This is still a very large file, considering typical corporate e-mail accounts only allow sending and receiving a maximum of 10 MB files.

Double-precision floating numbers are clearly still too inefficient. The size of data can be further reduced to 5 GB by using fixed-point integer representation (i.e., casting floating point S-parameter real/imaginary data) ($-1\cdots+1$) into $-32768\cdots+32767$ discrete integers such that it can be directly stored as a 16-bit (2 byte) signed integer.

Fixed-point representation only works for passive networks because the maximum S_{21} is bounded to 1 (0 dB). For active networks, such as amplifiers, a floating-point representation is required as the value of S_{21} has no upper bound. The smallest IEE754 standardized floating-point number is the half-float (2 bytes, 16 bits), which is the same number of bits as the fixed-point integer representation, but with less precision as 5 bits are used for the exponent.

7.3.4 IMPEDANCE TUNER COMPACT MODEL

Data reduction by limiting the precision of stored data cannot shrink the raw data enough for a large number of tuning states. A good alternative is to fit the measured data to a model and only store the model coefficients. Ideally, the compact model would have such a small memory footprint that it could be stored in cellular device memory and potentially used for a feed-forward real-time optimization algorithm. This compact model should be able to represent the intrinsic correlations between the input variables (DTC values) and output variables (complex S-parameters) without being dependent on the precision (number of bits) of the inputs but only depend on the degrees of freedom (number of tunable elements).

An example derivation of such optimal fitting function is presented later for the bandpass 3D tuning network shown in Figure 7.4. Every element has a series resistance associated with it to model dissipative losses. The equation for input impedance Z_{IN} (Equation 7.5) is derived using traditional linear circuit analysis techniques.

Input impedance for 3D tuning network shown in Figure 7.4 is

$$Z_{in} = \cfrac{1}{\cfrac{1}{R_{L1}+j\omega L_1} + \cfrac{1}{R_{C1}-j\cfrac{1}{\omega C_1}} \cdot 1 - \cfrac{1}{(R_{L2}+j\omega L_2)+\left(R_{C2}-j\cfrac{1}{\omega C_2}\right)+\cfrac{1}{\cfrac{1}{R_{L3}+j\omega L_3}+\cfrac{1}{R_{C3}-j\cfrac{1}{\omega C_3}}+\cfrac{1}{Z_0}}}}$$

(7.5)

The input reflection coefficient (S_{11}) can be calculated from input impedance using.

$$S_{11}=(Z_{in}-Z_0)/(Z_{in}+Z_0) \tag{7.6}$$

Solving S_{11} directly results in an extremely complicated function with too many terms to reproduce here. However, for the purposes of this analysis a numerical solution will suffice, which is derived by assuming that all DTCs follow a linear capacitance relationship $C_n=A_n{}^*x_n+B_n$ and all scalar terms (inductor values L_n, dissipative resistances R_{Ln} and R_{Cn}, frequency (w in radians), and DTC slope A_n and offset B_n) are set to a value of 1. Substituting these values into Equation 7.5, and using Equation 7.6 to calculate the numerical input reflection coefficient S_{11} of network shown in Figure 7.4 is

$$S_{11} = \frac{\begin{aligned}&[(4-j14)x_1+(0-j8)x_2+(4-j6)x_3+(8-j14)x_1x_2+(10-j8)x_1x_3\\&+(4-j6)x_2x_3+(13-j6)x_1x_2x_3+(0-j8)]\end{aligned}}{\begin{aligned}&[(-24+j34)x_1+(-16+j48)x_2+(-24+j34)x_3+(-36+j26)x_1x_2\\&+(-34+j12)x_1x_3+(-36+j26)x_2x_3+(-39+j0)x_1x_2x_3+(0+j48)]\end{aligned}} \tag{7.7}$$

Closer analysis of Equation 7.7 reveals that input reflection coefficient S_{11} can be modeled by a multivariate rational function having denominator and numerator complex polynomial terms that contain all unique combinations (e.g., x_1, x_2, x_3, x_1x_2, x_1x_3, x_2x_3, $x_1x_2x_3$) of products of the input variables and a complex constant, but no square or cubic terms. These can be easily enumerated by treating the combinations as all binary values of dimension n ($0–2^n$). For example, 3D tuner has eight unique combinations of input variables $[x_1, x_2, x_3]=[000, 001, 010, 011,...,111]$, where 1 means that input variable exists and 0 means it does not, and 000 refers to the constant.

Similar analysis can be applied to all of the other S-parameters to derive a generic 3D fitting function as shown in Equation 7.8.

Generic complex fitting function for all S-parameters of network shown in Figure 7.4 is

$$S_{ij} = \frac{v_1x_1+v_2x_2+v_3x_3+v_4x_1x_2+v_5x_1x_3+v_6x_2x_3+v_7x_1x_2x_3+v_8}{v_9x_1+v_{10}x_2+v_{11}x_3+v_{12}x_1x_2+v_{13}x_1x_3+v_{14}x_2x_3+v_{15}x_1x_2x_3+v_{16}},$$

$$v_n \in \mathbb{C}, \quad x_n \in \mathbb{Z} \tag{7.8}$$

TABLE 7.1

Examples of Complex Coefficients Fitted to Measured 3D Data at 1.0 GHz

Coefficients	S_{11}	S_{12}	S_{21}	S_{22}
v_1	$0.1006 - 0.7918i$	$0.0028 - 0.0054i$	$0.0028 - 0.0054i$	$-0.7150 + 0.2131i$
v_2	$-1.5556 - 1.6865i$	$-0.8548 + 1.9055i$	$-0.8548 + 1.9055i$	$-1.8592 - 1.2029i$
v_3	$-0.5650 + 0.2025i$	$0.0004 - 0.0009i$	$0.0004 - 0.0009i$	$0.0985 - 0.6019i$
v_4	$0.0730 + 0.1356i$	$0.0004 - 0.0014i$	$0.0004 - 0.0014i$	$0.1439 + 0.0327i$
v_5	$0.0467 + 0.0433i$	$0.0001 - 0.0001i$	$0.0001 - 0.0001i$	$0.0518 + 0.0328i$
v_6	$0.1303 + 0.0593i$	$-0.0003 - 0.0000i$	$-0.0003 - 0.0000i$	$0.0921 + 0.1058i$
v_7	$-0.0040 - 0.0033i$	$0.0000 - 0.0000i$	$0.0000 - 0.0000i$	$-0.0044 - 0.0025i$
v_8	$6.0340 + 4.1172i$	$-1.1136 + 2.2268i$	$-1.1136 + 2.2268i$	$4.5742 + 5.5984i$
v_9	$-0.1486 + 0.7979i$	$0.1399 + 0.9053i$	$0.1399 + 0.9053i$	$-0.0151 + 0.7800i$
v_{10}	$-0.9739 + 2.8708i$	$0.0132 + 3.4458i$	$0.0132 + 3.4458i$	$-0.4980 + 2.9187i$
v_{11}	$-0.2579 + 0.5914i$	$-0.0564 + 0.7325i$	$-0.0564 + 0.7325i$	$-0.1531 + 0.6052i$
v_{12}	$-0.0667 - 0.1415i$	$-0.1236 - 0.1270i$	$-0.1236 - 0.1270i$	$-0.0853 - 0.1253i$
v_{13}	$-0.0449 - 0.0467i$	$-0.0655 - 0.0334i$	$-0.0655 - 0.0334i$	$-0.0498 - 0.0379i$
v_{14}	$-0.0666 - 0.1307i$	$-0.1197 - 0.1164i$	$-0.1197 - 0.1164i$	$-0.0835 - 0.1156i$
v_{15}	$0.0040 + 0.0035i$	$0.0055 + 0.0024i$	$0.0055 + 0.0024i$	$0.0043 + 0.0028i$
v_{16}	$8.0640 - 1.5184i$	$8.0375 - 4.6416i$	$8.0375 - 4.6416i$	$7.3763 - 2.7998i$

In fact, similar function can be shown to apply to any n-dimensional impedance tuning network requiring a total of 2^{n+1} complex coefficients (8, 16, and 32 complex coefficients for 2D, 3D, and 4D tuners, respectively). To demonstrate the effectiveness of this method, the coefficients $v_1...v_{16}$ were calculated from measured S-parameter data set of a 3D 4-bit tuner, using Levenberg–Marquardt method for nonlinear least squares approximation. The coefficients are shown in Table 7.1.

Comparison between measured and modeled data is shown in Figure 7.5, demonstrating remarkable fitting accuracy across all tuning states. The significant feature is that the amount of data has been reduced from $2^{(3*4)} = 4096$ raw complex numbers per S-parameter and frequency point, to only 16 complex numbers, regardless of the number of bits for every DTC. For a 3D 4-bit tuner this represents a 256-fold reduction in the amount of data. Compared with traditional S2P files stored in ASCII format, this method results in a 4096 times smaller file size (0.15 GB versus 18.8 kB) assuming complex coefficients for 300 frequency points are stored using 16-bit signed integer format.

Given that measured data are often quite noisy with narrow resonances and other non-idealities, it is preferred to store the coefficients for every frequency point rather than trying to further reduce the memory footprint by fitting the complex coefficients across frequency. However, if desired, the frequency relationship can be implemented by fitting the complex coefficients using a complex vector fitting method.

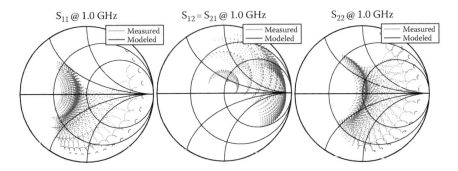

FIGURE 7.5   Measured versus modeled *S*-parameter data for a 3D 4-bit impedance tuner at 0.1 GHz.

7.3.5 IMPEDANCE TUNER VISUALIZATION

Being able to visualize behavior of tunable matching networks is critically important for optimizing their performance for a given application. However, given the sheer number of tuning states in typical impedance tuners (on the order of 10^3–10^9), a custom analysis software (written in mathematical analysis software such as MATLAB®) is often a better choice than trying to use generic circuit simulators (e.g., ADS). Several novel analysis techniques are presented in the following.

7.3.6 SMITH CHART CONSTELLATIONS

The Smith chart constellations of S_{11} and S_{22} are the most common way of visualizing the impedance tuning range of a tunable network. The spread of the points on the Smith chart indicates the impedance coverage of the network. Generally, impedance points scattered evenly and over a large percentage area of the Smith chart implies a high-gamma tuner (a tuner capable of matching large reflection coefficients and VSWR conditions). For lossless networks, the conjugate of S_{11} (complex reflection coefficient seen looking into Port1 when Port2 terminated to ZL, and vice versa) shows the exact source impedance that can be perfectly matched. However, for multidimensional tuning networks the Smith chart constellations are often cluttered, have multiple impedance points with different dissipative loss on top of each other and, thus, give only little insight into the behavior of the circuit.

7.3.7 SMITH CHART SURFACES

A very helpful aid in visualizing tuning networks is to plot reflection coefficients as surfaces stretched over the Smith chart. For a 2D tuner, the points on the grid are defined by two variables (e.g., DTC1 as x axis, DTC2 as y axis), as shown in Figure 7.6b. A checkerboard coloring is used to further reveal the built-in structure in the data. From a surface plot, a constellation fold-over is obvious, that is, part of the constellation is on top of another part, but it is not directly evident from the "flat"

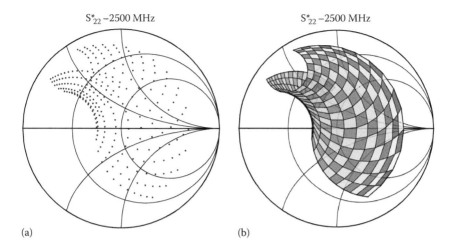

FIGURE 7.6 Smith chart constellation (a) and surface (b).

regular constellation. Care should be taken to properly select which DTC is mapped to which axis such that the resulting graph conveys the information properly. For higher dimensional tuners, some dimensions will need to be kept fixed while others are mapped to a surface.

7.3.8 TRANSDUCER GAIN CONTOURS

To characterize insertion loss of the matching network, the S_{21} can only be used if both ports are same as characteristic impedance (typically 50 Ω). Since the whole purpose of tuner is to match non-50 Ω source and load impedances, transducer gain equations (Equation 7.9) must be used to calculate the effective insertion loss between complex port impedances.

The transducer gain equation for determining matching loss for complex source and load impedances is

$$G_T = 10\log_{10}\left[\frac{(1-|\Gamma_G|^2)|S_{21}|^2(1-|\Gamma_L|^2)}{|(1-S_{11}\Gamma_G)(1-S_{22}\Gamma_L)-S_{12}S_{21}\Gamma_G\Gamma_L|^2}\right]$$

$$\Gamma_G = \frac{Z_G-Z_0}{Z_G+Z_0}, \ \Gamma_L = \frac{Z_L-Z_0}{Z_L+Z_0}$$

(7.9)

Contours of transducer gain drawn on a Smith chart are immensely useful in tunable matching network analysis and optimization. Typical Smith chart constellations of S_{11} and S_{22} only show points that can be conjugate matched, but they do not include any information on the dissipative loss of that particular tuning state or information about areas of Smith chart that are outside of the constellation. The transducer gain

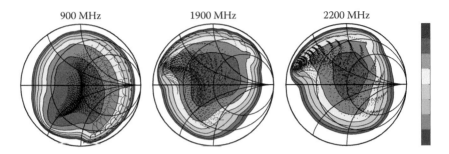

FIGURE 7.7   Load transducer gain contours and conjugate of S_{22}.

contour, on the other hand, shows the area of the Smith chart that can be matched to the source with a given loss performance. For best performance, the load impedance of, for example, antenna or filter input impedance should be pre-matched or otherwise placed inside the low-loss contour. The difference between conjugate of S_{22} constellation plot and load transducer gain contours is illustrated in Figure 7.7. The coverage areas of the S_{22} constellation do not always overlap with reasonable (e.g., −1 dB) loss contour.

Transducer gain contours can be created by a mathematical load-pull technique using S-parameter data, which is valid for linear networks such as impedance tuners. That is, represent every complex load impedance reflection coefficient on the Smith chart as a grid of complex values (e.g., values −1 to 1 at 0.01 steps for both real and imaginary parts for a total of 40,401 complex numbers). Then for every load reflection coefficient and every tuning state (e.g., 1,048,576 states for a 4D 5-bit tuner) calculate transducer gain using Equation 7.9, and then pick the tuning state with highest transducer gain. Now this data set can be used to draw contours of transducer gain overlaid on top of the Smith chart. This technique works but given the brute force nature it gets very inefficient for large number of tuning states or fine load reflection coefficient grid. In this example, 40 billion transducer gain function evaluations (complex equation so computationally very expensive) would be required for every frequency point.

A novel alternative technique is presented later that is several orders of magnitude less computationally complex. It is based on calculating a transducer gain circle (for a given contour level, e.g., −1.0 dB) for every tuning state and then graphically tracing the outline of all the G_T circles to create the contour (Figure 7.8). The required calculations are shown in Equation 7.10, where g is the normalized gain (relative to peak S_{21}) given contour level P (in dB), c is the center of the circle (in complex coordinates of reflection coefficient and r is the circle radius as an absolute number).

Equations for calculating transducer gain circles are

$$g = \frac{10^{P/10}}{|S_{21}|^2 \left(1/\left(1-|S_{22}|^2\right)\right)} \qquad c = \frac{gS_{22}^*}{1-(1-g)|S_{22}|^2} \qquad r = \frac{\sqrt{1-g}\left(1-|S_{22}|^2\right)}{1-(1-g)|S_{22}|^2} \qquad (7.10)$$

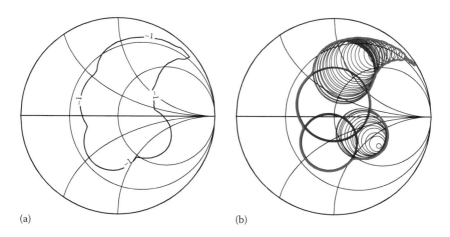

FIGURE 7.8   Traditional brute-force transducer gain contour method (a) compared with a gain circle boundary method (b).

The use of transducer gain circles removes the need to evaluate the transducer gain function at every load reflection coefficient value, as the gain circle already contains that information. In the earlier example, this reduces number of function evaluations by a factor of over 40,000. An efficient sweep-line algorithm or alpha boundary can then be used to trace the boundary of the gain circles to create the contour in logarithmic time.

The transducer gain circle boundary method is a significant improvement over the brute-force method but still inefficient as the number of function evaluations is proportional to number of tuning states. This becomes very impractical for tuners that have a large number of bits per DTC (e.g., 6–8 bit), because number of unique tuning states grows exponentially with the number of bits per DTC. To completely eliminate the dependency on number of bits per DTC, the transducer gain circle boundary method can be combined with tuner compact model equations and coefficients. This allows that transducer gain contours to be analytically calculated directly from the complex rational function, without ever having to evaluate the complex function at discrete DTC values to create discrete S-parameter data. This would provide practically instantaneous transducer gain contour calculation from the tuner compact model coefficients, for any complexity.

7.3.9   DENSITY PLOTS FOR VISUALIZING LARGE AMOUNTS OF DATA

For higher-dimensional tuners with significant number of states (>1 e4) it is impractical to plot every single tuning state as its own line plot. There are too many lines on top of each other, which tends to mask any underlying groupings in the data. An alternative way to display the data are to create bins and use a histogram function to calculate number of points going through each bin, and then display a bitmap of the bins. This reveals areas where most of the data points are concentrated. Figure 7.9 shows comparison between regular line plots and density transducer gain for a tuning network with several thousands of tuning states.

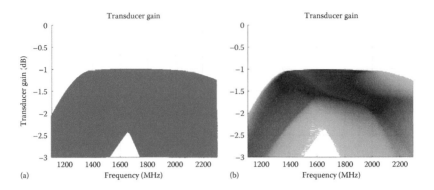

FIGURE 7.9   Line plot (a) versus density plot (b) for a large number of tuning states.

7.4 CONCLUSIONS

Over recent years, the complexity of cellular devices has grown significantly with the exponential growth in mobile data demand, which has had a significant impact to the complexity of cellular RF front end. What once was a simple single-antenna single-band transmit and receive duplexing solution with no active RF switching has turned into a complex RF front end supporting 10+ FDD/TDD bands, 16+ RF switch throws, multiple simultaneous carriers, reconfigurable multimode and multiband power amplifiers, and impedance and antenna tuning. Optimization of the performance, size, and cost has practically displaced GaAs in the switching portions of the RF front end by enabling monolithic integration of switching, tuning, and amplification functions using RF-CMOS SOI. While silicon varactor, BST, and MEMS-based alternatives exist for tuning applications, switched CMOS capacitors exhibit a unique trade-off between RF performance, simplicity, and monolithic integration. To the first order the RF performance is dominated by the $R_{ON}C_{OFF}$ figure-of-merit of the CMOS switch process, but the peak voltage, impedance tuning range, and dissipative loss behavior is significantly impacted by second order parasitic effects. A careful optimization of all parameters is required to achieve best performance. While RF tuning improves the radiated performance of the mobile device, it relies on being able to select the one optimal tuning state out of thousands or millions of possible candidates. Traditional RF circuit simulation, measurement, and optimization techniques have to be replaced by size and speed optimized data structures, analysis, and measurement techniques.

REFERENCES

1. Ranta, T., Ella, J., and Pohjonen, H., Antenna switch linearity requirements for GSM/WCDMA mobile phone front-ends, *8th European Conference on Wireless Technology Proceedings*, Paris, France, pp. 23–26, October 3–4, 2005.
2. Ranta, T. and Novak, R., New tunable technology for mobile-TV antennas, *MWJ*, 2008.
3. Ranta, T. and Novak, R., Antenna tuning approach aids cellular handsets, *MW& RF*, 2008.

4. Ranta, T. and Novak, R., Improve mobile handset antenna performance with new tuning techniques, *EE Times*, 2008.
5. Ranta, T., Pilgrim, D., and Whatley, R., RF front end adapts for increased mobile data demand, *EE Times*, 2010.
6. Whatley, R., Ranta, T., and Kelly, D., RF front-end tunability for LTE handset applications, *Compound Semiconductor Integrated Circuit Symposium* (*CSICS*), 2010 IEEE, pp. 1–4, October 3–6, 2010.
7. Whatley, R., Ranta, T., and Kelly, D., CMOS based tunable matching networks for cellular handset applications, *Microwave Symposium Digest* (*MTT*), 2011 IEEE MTT-S International, pp. 1–4, June 5–10, 2011.
8. Baxter, B., Ranta, T., Facchini, M., Jung, D., and Kelly, D., The state-of-the-art in silicon-on-sapphire components for antenna tuning, *Microwave Symposium Digest (IMS)*, 2013 IEEE MTT-S International, pp. 1–4, June 2–7, 2013.
9. Cheng, C., Facchini, M., Ranta, T., and Whatley, R., High performance 1.8–2.4 GHz phase shifter using silicon-on-sapphire digitally tunable capacitors, *Microwave Symposium Digest (IMS)*, 2013 IEEE MTT-S International, pp. 1–3, June 2–7, 2013.
10. Ranta, T., Whatley, R., Cheng, C., and Facchini, M., Next-generation CMOS-on-insulator multi-element network for broadband antenna tuning, *Microwave Conference (EuMC)*, 2013 European, pp. 1567–1570, October 6–10, 2013.
11. Sekar, V., Cheng, C., Whatley, R., Zeng, C., Genc, A., Ranta, T., and Rotella, F., Comparison of substrate effects in sapphire, trap-rich and high resistivity silicon substrates for RF-SOI applications, *2015 IEEE 15th Topical Meeting on Silicon Monolithic Integrated Circuits in RF Systems (SiRF)*, pp. 37–39, January 26–28, 2015.

8 Handset Antenna Tuning Using BST

Paul McIntosh

CONTENTS

8.1   OVERVIEW

This chapter will review barium strontium titanate (BST)-based handset antenna tuning, which has become increasingly important for the today's multiband, multimode, sleek, and ubiquitous smartphone. As the smartphone screen size has increased, the battery size has increased to support larger and brighter screens along with more and more functionality; hence, the volume available for the antenna(s) has been reduced.

This chapter will review feedpoint antenna tuning that uses BST-based tunable capacitors in the impedance matching network at the feedpoint of the antenna, and this chapter will also present a design approach and case study. Antenna feedpoint tuning is becoming increasingly common and is conceptually simple, as it replaces the fixed match that most handsets employ with a circuit that uses tunable components.

8.2   ANTENNA MATCHING AND THE TRANSITION TO TUNABLE MATCHING NETWORKS

Mobile phone antennas have been typically designed to operate into a 50 Ω load impedance, but in practice, achieving this directly from the antenna is very difficult. The wide frequency range of a modern handset's antenna means that a matching network is often needed to ensure that the antenna can transfer the maximum power into the receiver or radiate the maximum power from the transmitter. A matching network made of non-tunable components, has been the only way of building handset antenna

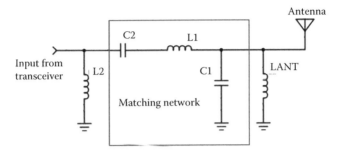

FIGURE 8.1 Example of an antenna matching network using fixed elements.

matching networks until recently. However, with the proliferation of bands that are now used, covering frequencies from as low as 698 MHz to as high as 2.7 GHz routinely and even as high as 3.5 GHz,[1] the performance of a matching network comprised of fixed components can significantly limit the overall performance of the handset. An example of a matching network using fixed components is shown in Figure 8.1.

A matching network designer using fixed components must make compromises during the synthesis by trading off performance at some frequencies in order to achieve acceptable performance over the whole band. Practically, this is achieved using a range of techniques, including hand adjustments and optimization using a vector network analyzer (VNA), impedance matching design using the Smith chart, and computer-aided design (CAD) using software packages such as Optenni Lab,[2] Advanced Design System (ADS),[3] and Microwave Office.[4] Clearly, once selected the performance is fixed and cannot be adjusted at all to accommodate any changes in the antenna's scattering parameters (S-parameters) due to external loading. Figure 8.2 illustrates the frequency response of a fixed matching network where the performance has been centered at 800 MHz. The performance cannot be adjusted to improve the voltage standing wave ratio (VSWR) at either 700 or 900 MHz, nor can the behavior be modified to compensate for the loading effects of the hand or head that occur during real-world use.

Tunable antenna matching networks, commonly referred to as antenna tuners, are not new structures, having been commonly used by amateur radio enthusiasts for many years. In the amateur radio application, tunable matching networks have been realized using air variable capacitors and switched inductors. In the case of cellular handsets though, the lack of suitably high-performance tunable radio frequency (RF) components has limited the use of antenna tuners. There has been a

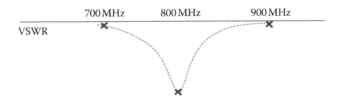

FIGURE 8.2 Fixed matching network response.

wide range of work published on antenna tuners, including circuits that use BST-based capacitors,[5] switched capacitor structures,[6] and varactor diodes,[7] but until recently, antenna tuners were mostly confined to the published literature. The published results are illustrative of the potential capabilities but do not often address performance in real handsets.

However, three important factors have occurred together leading to antenna tuners now becoming a standard feature in handsets. These are the emergence of high-performance tunable elements, such as BST-based tunable capacitors, the increasing frequency range used for handsets, particularly the extension down to 698 MHz band for fourth-generation Long Term Evolution services (4 G LTE), and the consumer demand for slim handsets where little volume is available for the internal antenna. These factors combined have led to handset design teams embracing antenna tuners in an ever increasing number of commercially available handsets.

The matching network shown previously in Figure 8.1 can be modified to become a tunable network by simply replacing the fixed capacitors with tunable capacitors, as shown in Figure 8.3. Conceptually, this is a simple change with BST-based tunable capacitors, or Passive Tunable Integrated Circuits (PTICs™), being used along with the addition of a bias controller. This bias controller is used to supply the required bias voltage to the PTIC and also interfaces with the handset baseband IC, which controls the operation of the phone including the entire RF chain.

As a result of the replacement of the fixed capacitors with PTICs, the network response, shown in Figure 8.4, is now variable and can be adjusted to compensate

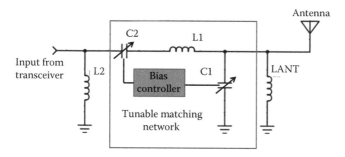

FIGURE 8.3   Tunable antenna matching network.

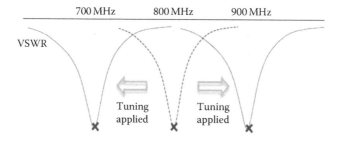

FIGURE 8.4   PTIC-based tunable network response.

for the effect of hand or head loading as well as being tuned to the frequency of operation. It is also important to note that the frequency response of the BST tunable capacitor–based network can be tuned in an analog fashion, due to the continuously variable nature of the PTICs. This, along with the fact that PTICs are analogous to fixed caps meaning that they can be used as single devices in shunt or series, allows for tremendous flexibility in the matching network.

8.3   ANTENNA MEASUREMENTS FOR TUNABLE MATCHING NETWORK DESIGN

Accurate measurement of the antenna S-parameters (1-port measurement) is critical since it provides the correct terminating impedances for the network across the frequency range and also establishes the reference plane where the matching network will be placed. Typically, antenna measurements will be done many times during the handset development process. This is due to the increasing maturity of the industrial design (ID) as the design project progresses and the associated antenna changes that will be required to accommodate the changes in the ID. The antenna S-parameters will need to be measured under a range of use cases if these are to be considered in the tunable matching network design as well. The baseline use case is known as freespace and is a measurement of the antenna without any impairment. No additional fixturing is required for freespace measurements and it only requires that the handset be mounted in such a way that nothing is interfering with the antenna.

The handset antenna measurements require the installation of an RF cable into the unit that will connect to the antenna feedpoint at the exact location of the first matching network element when looking from the antenna. This is shown in Figure 8.5, where the coaxial cable is terminated at the antenna spring contacts. The location of the cable is very important since the S-parameters will be referenced to this plane. If the cable is not placed at the correct reference plane, there will be an

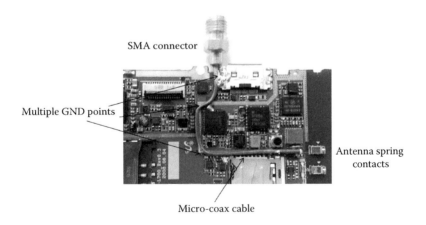

FIGURE 8.5   Example antenna measurement cable installation.

additional phase shift and loss that is unaccounted for in the design, which will have a detrimental effect on the performance of the network. It is recommended that a micro-coaxial semirigid type cable be used, which can be easily routed within the phone, and the outer jacket of the cable can be soldered to suitable grounding points along its length, as shown in Figure 8.5.

The cable should be routed away from the antenna to reduce its impact on the antenna behavior and the S-parameter measurements. The measurements can be conducted using an automated VNA with port extension capabilities such as those offered by Rhode and Schwarz[8] or Keysight.[9] Using the port extension function with the cable either grounded or open at the antenna feedpoint, that is, the spring contact in this particular example, the effect of the cable will be accounted for. The phone is now ready for antenna S-parameter characterization in the freespace use case and other conditions as required. Figure 8.6 shows the type of phantom that can be used in the characterization of the antenna in other use cases, to replicate head loading and hand-loading conditions.

It is also worth noting that this same handset, with the cable installed, can be used for antenna efficiency measurements. These measurements allow the radiation performance, known as radiation efficiency of the antenna, to be established and when combined with the two-port analysis of the matching network, gives the total performance of the antenna subsystem. Similar measurements can be made across the entire frequency range of the handset and when cascaded with known amplifier output power and/or receiver sensitivity in each band provides a good estimate of the performance of the handset.

The results of the antenna measurement can be plotted on the Smith chart and, along with the efficiency measurements, provide a good overall picture of the antenna behavior. The S-parameters for an example antenna measured under three different use cases—freespace, phantom hand (PH) grip, and for contrast a human hand grip are shown in Figure 8.7. The S-parameters for each of these use cases can be used in the design of a tunable matching circuit if desired. Figure 8.7 shows that while a PH does provide some loading and causes a shift in the S-parameters, the effect of the human hand is considerably greater. This illustrates two very important points. First, characterization under real grip conditions, when the hand use case is being

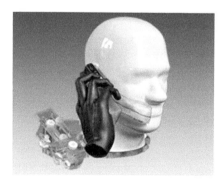
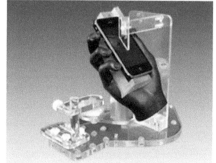

FIGURE 8.6   Head-and-hand phantoms.

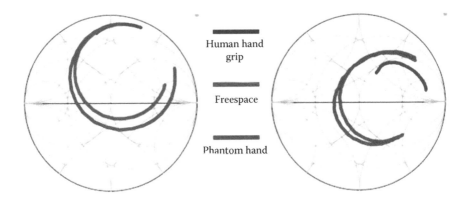

FIGURE 8.7   Example antenna *S*-parameters under multiple use cases (low band on the left, high band on the right).

considered, is important to establish true loading effects. Second, the impact of the user's hand can be quite large. In the case of a fixed-component matching network, the antenna engineer will have to select a use case to design for. This may lead to them compromising the other use case(s), or if they attempt to design for multiple use cases, it is likely that they will end up compromising on performance over all operating conditions. The benefit of a tunable matching network, where the network response can be tuned to accommodate the operation of the handset under a variety of use cases, is conceptually clear.

8.4   TUNABLE MATCHING NETWORK TOPOLOGIES AND CIRCUIT OPTIMIZATION

A tunable matching network can be composed of a number of different stages with one, two, or even three tunable capacitors. The greater the number of stages, the more flexibility there is in the network design, but this has to be traded off against the insertion loss of the network. Adding more components adds more loss. Two of the most common networks are the Pi and T circuits shown in Figures 8.8 and 8.9, respectively.[10,11] In Figures 8.8 and 8.9, there are three matching components that are represented by generalized impedances. In the case of a tunable matching network, these would be replaced by PTICs, which would provide the tunable element paired with an inductor, as shown in Figure 8.10. By pairing the PTIC with an inductor, a greater range of impedances can be synthesized than if a capacitor alone is used, and there are more degrees of freedom available to the designer during the optimization step. These generalized cases use three tunable elements although this could be reduced to two or even a single element in order to ensure the right balance of improved matching performance, insertion loss of the network, and bandwidth.

In the case of a particular antenna, it is important to be able to determine which circuit or subcircuit has the best performance. Subcircuits are versions of the general three-element case that have only two or even one PTIC. A two PTIC structure may be

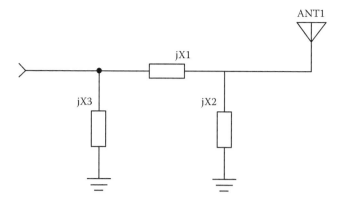

FIGURE 8.8 Pi matching topology.

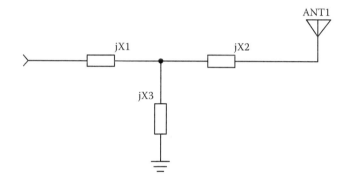

FIGURE 8.9 T matching topology.

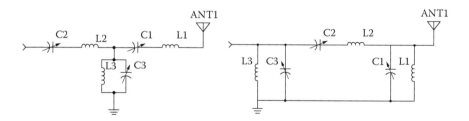

FIGURE 8.10 Three PTIC Pi and T matching circuits.

suitable when the antenna is not heavily mismatched (<4 dB mismatch). In this case, the introduction of a third stage would cause additional loss for very little additional improvement in matching coverage. In the case of a strongly mismatched antenna this situation may be reversed; the additional loss of a third stage could be tolerated in order to achieve better matching performance. Useful examples of two PTIC matching circuits are shown, one previously in Figure 8.3 and one that follows in Figure 8.11.

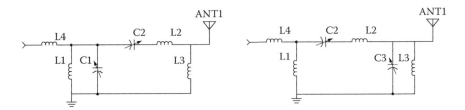

FIGURE 8.11   Two PTIC matching topologies (based on a Pi network).

Since there are several versions of matching circuit that can be used, a computer-aided simulation is needed in order to optimize and compare the individual circuits. The general process is to analyze each matching circuit, or tuner, and then determine the performance of the system containing that matching circuit compared with the performance of direct connection between the antenna and the system impedance, typically 50 Ω. This will allow the benefit of the tuner to be established. Comparing the relative benefit will allow the best circuit to be selected. The improvement that the tuner provides can be summarized as shown in Equation 8.1 from Gu and Morris.[12] This is a calculation of the relative transducer gain (RTG) and is a measure of the improvement that the tuner yields when compared to the performance without a tuner:

$$RTG = GT_{Tuner} - GT_{Without\ Tuner} \qquad (8.1)$$

When the power wave definition of S-parameters is used with arbitrary complex input and output termination impedances, then the transducer gain can be represented as shown in Equation 8.2.[13] A useful description of the application of power waves to the design of conjugate matching impedances can also be found in Rahola:[14]

$$GT = \left|S_{21}\right|^2 \qquad (8.2)$$

In the analysis of a given matching circuit, the return loss or input match, are often examined and optimized either by hand on a VNA, or using a circuit simulator. However, in the case of a real circuit with loss from the matching elements, this approach will not lead to the optimum performance as it ignores the dissipative loss of the matching network components. Instead, the transducer power gain for each circuit should be evaluated to ensure that the circuit has the optimum performance.[15] The loss of the matching network is composed of the dissipative loss of the components, such as inductor and capacitor Q, and the mismatch loss due to non-perfect impedance matching. The optimization of each circuit should then focus on targeting maximum $|S_{21}|^2$ in each band of operation.[16] By using this parameter as the main optimization variable, it will ensure that circuits in which the return loss is better, but the total loss is higher are avoided. Matching networks with good return loss results do not necessarily yield optimum performance since the goal of the matching network is to achieve maximum power transfer to and from the antenna. By computing the transducer power gain, the best schematic values can be selected.

The optimization process will need to set all of the matching components as variables and use $|S_{21}|^2$ as the optimization criteria. This optimization process will need to include all of the frequencies that will be used in the handset if those frequencies are for duplex or simplex operation. In the case of time division duplex systems (TDD) such as GSM or TDD-LTE, the matching network settings can be optimized for each transmit and receive frequency. For example, in the GSM 900 band, the transmit frequency of 880 MHz can be optimized independently from the receive frequency of 925 MHz since transmit and receive operations occur in different time slots. In the case of a PTIC-based matching network, the PTICs can be tuned to particular values for the transmit operation and then changed for the receive operation. This reconfiguration will ensure that the optimal performance is obtained at both frequencies. In the case of frequency division duplex (FDD) systems such as FDD-LTE or UMTS, the transmit and receive frequencies are paired. The matching network must operate simultaneously at both frequencies. For example, in Band 1 UMTS (WCDMA) the transmit frequency is 1920 MHz and the receive frequency is 2110 MHz. The optimization of the matching network must be carried out to accommodate the transmit/receive frequency spacing of nearly 200 MHz. The PTIC bias voltages that are used for the transmit frequency must be the same as those used for the receive frequency. The PTIC bias voltages can be adjusted for optimum performance when a different channel is used in the handset, since this means that a new transmit/receive frequency pair will be used.

The number of PTIC bias voltages used within a band can be as simple or as complex as desired and the decision on how many to use is part of the optimization process. In the case of FDD operation, there can be a single setting for each capacitor for each band. However, this may lead to suboptimal performance so a better approach is to subdivide a given band into two or three sets of subchannels. Using this approach and three subbands, the matching network will have a different set of PTIC bias voltages for the lower, middle, and upper third of the band. An example is shown in Table 8.1.

The use of a circuit simulator is required in order to optimize and compare the performance of each topology that is examined. There are a wide range of choices that can be used in which the designer can build their own simulation setup, such as AWR's Microwave Office simulator, Ansoft's Designer suite, Keysight's (formally Agilent) Advance Design System, and free packages, such as the open source Quite

TABLE 8.1
Example of Subband PTIC Voltages

Band	Mode	Channel Range (Tx/Rx)	PTIC Settings (Two PTIC Tuner)
I	UMTS FDD	9,612–9,703/10,562–10,653	V1, V2
I	UMTS FDD	9,704–9,796/10,654–10,745	V3, V4
I	UMTS FDD	9,797–9,888/10,747–10,838	V5, V6

Universal Circuit Simulator (QUCS) and simulators specifically designed for matching network design such as Optenni Lab. In all cases, it is important to use accurate antenna S-parameters and realistic models of the fixed elements, such as inductors and capacitors as well as the PTICs. A data file of PTIC S-parameters can be used in order to reduce the burden on the designer to extract their own models. These are available from ON Semiconductor Corp. For the inductors, the manufacturer, Murata or Coilcraft, for example, can provide accurate device models. Optenni Lab offers a very simple interface to the simulator where the user can import PTIC S-parameters and specify the number of elements required in the matching network. After importing the antenna S-parameters and specifying the frequency bands that the network has to cover, the simulator will automatically optimize the topology with the selected number of elements.

The simplest optimization process will use a single set of antenna S-parameters, such as the freespace use case, but additional use cases can also be included to ensure that the matching network performs adequately under a range of different conditions. The optimization process can also be constrained so that component values, such as inductors, remain within a usable range for handset matching circuit application. By accurately modeling the component losses, the overall performance of the antenna system (antenna and matching network) can be determined. The total efficiency can be obtained by cascading the antenna radiation efficiency with the matching network transducer gain when the network is terminated by the antenna's S-parameters.

8.5 IMPLEMENTATION OF PTIC ANTENNA TUNERS IN A MOBILE PHONE

The typical design of an antenna prior to the introduction of tunable elements would have been focused on a ensuring that the antenna was as well matched as possible into 50 Ω across the entire frequency range of operation or could be easily matched with the application of a matching network. However, with the availability of PTICs that can be used in tunable matching networks, the antenna can be designed with a focus on ensuring the highest possible radiation efficiency even though the antenna may be mismatched into 50 Ω. The total antenna efficiency is a combination of the antenna radiation efficiency and mismatch loss and shows the overall performance of the antenna. In a fixed matching system, a highly mismatched, but efficient antenna would be problematic, as the benefit of the high radiation efficiency could be outweighed by the mismatch loss, which may be unacceptably large at many frequencies. However, in the case of a tunable network, the matching network can be adjusted for each different frequency or frequency pair (duplex tuning case). This then addresses the issue of a highly efficient, but mismatched antenna. The use of an antenna tuner allows the antenna designer the freedom to design for maximum radiation efficiency, free from the constraints of a fixed matching network.

The PTIC is a uniquely simple device from a control perspective with the variation in capacitance being controlled by the DC bias. This means that multiple devices can be used within a phone and all the bias and interfacing to the baseband control system can be centralized. ON Semiconductor offers multiple versions of the bias and control IC,[17] which have differing numbers of bias outputs from 2 to 6.

Depending on how many PTICs are being used within a phone, a cost-effective implementation can be achieved. Since all the control and digital interface circuitry is centralized, removing any duplication of digital functions, the solution can be kept to the most efficient implementation. A good example of this is when PTICs are deployed in matching circuits used for both main and diversity antennas. The design of the diversity antenna can be just as difficult as the main antenna and, in some cases, more difficult due to even more challenging space constraints. If we consider an example where a single PTIC matching network is used for the diversity antenna and a two PTIC matching network is used for the main antenna, both networks can easily be controlled using a three-output controller. There are minimal special requirements for the DC bias lines, making routing more straightforward. Since the digital and control functions are in a single IC, they can be kept away from the antennas to ensure there is no risk of any spurious noise affecting antenna performance.

In a more complex handset where there are two antennas both used for transmit and receive, the solution can be further scaled with minimal additional circuitry needed. In this case, three PTICs could be used in each matching circuit so a six output control IC is used. The same general approach is used for all designs with the central control interface communicating with the baseband processor and providing the DC bias control to each of the PTICs in the matching networks. As it can be seen, the use of small, DC biased tunable capacitors, combined with the PTIC controller allows for the support of many different types of architecture. These PTIC matching networks can be easily deployed at the port of each antenna that requires tuning.

8.6   EXAMPLE OF A PTIC TUNER DESIGN

In this section, we will analyze a design of a tuner for a PIFA style antenna that has to cover 700–960 and 1710–2700 MHz. The tuner S-parameters are examined, and a simple optimization for freespace operation has been carried out where the goal has been to maximize the transducer power gain of the network at multiple frequencies. The examination of the RTG, as shown in the previous section, shows the improvement that the PTIC network can provide. The benefit of being able to adjust the network response in order to match the frequency band of operation is also demonstrated.

A handset has been modified to include an RF cable so that S-parameters can be measured using a VNA. The antenna efficiency will also be measured as part of the design process but is not presented here as we are focused on the matching network design and reduction in mismatch loss of the antenna tuner in this section. The S-parameters of the antenna have been recorded and are plotted on the Smith charts shown in Figure 8.12. In the low band at the band edges, the reflection coefficient is close to the edge of the Smith chart, which will result in a high mismatch loss. As the frequency increases, the input impedance gets closer to the center of the Smith chart and 50 Ω. Around the middle of the low band, at approximately 880 MHz, the match improves and at this point the mismatch loss will be low, since the antenna is reasonably well matched. As the frequency increases further, the input match into

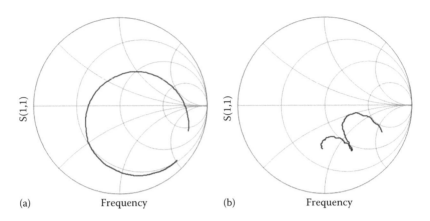

FIGURE 8.12   Antenna S-parameter—S_{11}: (a) 0.6–0.96 GHz and (b) 1.7–2.7 GHz.

50 Ω begins to degrade again, resulting in an increase in mismatch loss at the high-frequency band edge. The performance in the mid and high bands are an interesting counter to this since the antenna is better matched across these bands even though the mismatch loss is high at certain frequencies.

Figure 8.13 shows the return loss of the antenna in both frequency bands and the variation of the match into 50 Ω. Figure 8.14 shows the mismatch loss in the low, mid, and high bands. The antenna in the low band has ~6 dB of mismatch loss at 700 MHz. This is the lowest frequency of operation and coves LTE bands 12, 13, and 17. These are challenging bands to design for since the lower frequency requires an electrically larger antenna, but the available space in a modern handset is limited. The antenna is actually well matched in the high bands and suffers from little mismatch loss with the average loss being around 1 dB. However, at the highest frequency of operation, the mismatch loss does begin to increase. Given that this is single feed antenna, that is, one feedpoint covers the entire operating frequency range, the design of a matching network poses an interesting challenge. The matching network must be sufficiently low loss that the already good insertion loss and

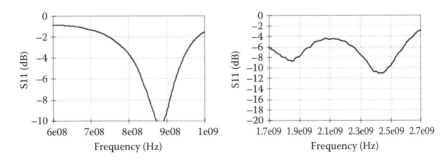

FIGURE 8.13   Antenna return loss.

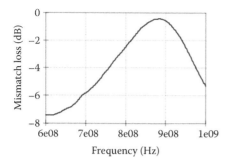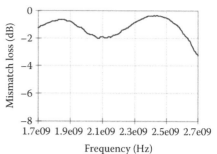

FIGURE 8.14 Antenna mismatch loss.

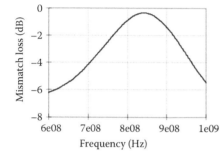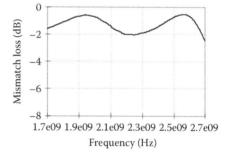

FIGURE 8.15 Antenna mismatch loss under BHHR loading conditions.

input match of the high band is not compromised too much, but still provide a good match at low frequencies where the antenna mismatch is high.

In this design example, the antenna behavior under a use case condition will also be considered. The antenna's response will alter under loading conditions caused by the user's head or hand and a tunable matching network can be adjusted to compensate for these effects. A fixed matching circuit does not have any flexibility so the designer will either design for a particular use case, typically freespace, or a compromise between freespace and the various loading conditions. In this simple example, the beside head use case will be considered. Figure 8.15 shows the mismatch performance of the antenna under the beside head hand right (BHHR) loading condition.

As can be seen in Figure 8.15, the loading effect of the beside head hand use case keeps the mismatch loss at 700 MHz to approximately 4 dB and at 960 MHz to more than 4 dB. Both of these are still significant mismatches and the high VSWR makes the design of the matching network important to keep the insertion loss as low as possible. Any loss in the matching network will reduce the output power of the phone and result in a direct reduction in the total radiated power (TRP) and total isotropic sensitivity (TIS). The high band has a very similar performance to freespace and is relatively unchanged by the impact of the PH. As expected, the loading effect is more prominent in the low bands.

The first step in designing the matching network is to set the frequency bands that will be used in the handset. There are a large number of bands that a handset needs to address and for the purpose of this design the following bands will be considered: bands 7, 12, 13, 17, and 40 for LTE operation and bands 1, 2, 3, 4, 5, and 8 for WCDMA operation. These bands will also cover the four GSM bands, GSM850, 900, 1800, and 1900. Operation in either LTE or WCDMA mode will require duplex settings for the PTICs, but the GSM bands can be optimized for either simplex operation, with separate settings for transmit and receive operation, or duplex operation. For simplicity in this example, the GSM bands are also optimized for duplex operation. The user will also need to decide if the bands will be subdivided at all. This provides finer control within a band and can yield improved performance. In this example, the bands will be divided into three with PTIC settings for the low, middle, and high channels. However, in practice, it may be that some of the tuner states are sufficient to cover an entire band.

The terminating impedances also need to be defined for the optimization. In the general case, the antenna impedance file(s) (touchstone format.s1p) are used as the load impedance(s) and the source impedance can either be 50 Ω or the result of load-pull measurements of the transceiver. Using load-pull measurements, if available, can be beneficial since the output of the transceiver when measured at input of the matching circuit may not be 50 Ω. In the case of source impedances that are varying over frequency, an additional .s1p file can be specified and used instead of 50 Ω. For example, the Optenni Lab software allows the user to import an impedance file for use as the source terminations. In this example, 50 Ω will be used for simplicity.

Finally, the goals of the simulation need to be established. As discussed previously, it is important to optimize the circuit values for maximum transducer gain, or minimum insertion loss, rather than return loss. Optimizing for return loss may result in a good, but lossy, match being selected that will not give the maximum power transfer. There may also be the need to weight the performance of a particular band, such as applying greater weight to the home network bands rather than the roaming bands, which are used only rarely. This can be achieved by having a different $|S_{21}|^2$ goal for a particular band, such as specifying $|S_{21}|^2 < -1$ dB for the home bands and $|S_{21}|^2 < -2$ dB for roaming bands. Once the design criteria have been established, the simulation/optimization of different topologies can be undertaken in a simulator of the designer's choice.

8.6.1 Circuit Analysis Results

Using the *S*-parameters as shown earlier in Figures 8.12 through 8.15, designs have been carried out for a two PTIC tunable matching circuit. The circuit topology used is shown in Figure 8.16 and, as discussed earlier, it has been optimized for 11 bands, each divided into 3 subbands. In this particular case, 50 Ω has been used as the reference impedance rather than load-pull targets, although load-pull impedances could also be used if available, as discussed previously. The resulting schematic values from the optimization are also shown.

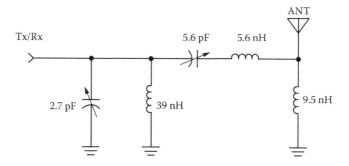

FIGURE 8.16 Antenna tuner for case study.

In this circuit design, the PTICs are simulated using *S*-parameter blocks, which represent the full performance of the devices over frequency and bias. The inductors are modeled with a *Q* of 60 at 1 GHz, which represents the performance of commonly available RF wirewound-type devices. The maximum capacitance value of the PTICs is specified at 2 V bias. The device can be continuously tuned from 2 to 24 V with a typical tuning range of 4.5:1. An example of the PTIC capacitance versus voltage characteristic is shown in Figure 8.17.

In order to understand the benefit of the tunable matching circuit, the performance can be examined in a couple of different ways using the RTG method as previously reviewed. In the first instance, the RTG can be calculated by comparing the results with the matching network in operation against no matching network. This will allow the improvement in mismatch loss (MML) to be clearly seen. Even though

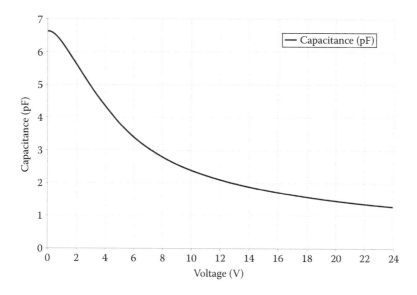

FIGURE 8.17 PTIC capacitance versus voltage characteristic.

TABLE 8.2

Tuner Design Case Study Data in the 700 MHz LTE Bands for the FS Condition

Frequency (MHz)	Band/Mode	Channel	MML with Tuner (dB)	MML without Tuner (dB)	Tuner Insertion Loss (dB)	Relative Transducer Gain (dB)
698	12/LTE	L-Tx	−0.4	−5.8	−1.6	4.2
716	12/LTE	H-Tx	0.0	−5.4	−1.0	4.4
728	12/LTE	L-Rx	−0.4	−5.1	−1.1	4.0
746	12/LTE	H-Rx	−0.8	−4.4	−1.6	2.8
777	13/LTE	L-Tx	−1.2	−3.3	−1.6	1.6
787	13/LTE	H-Tx	−1.1	−2.9	−1.5	1.4
746	13/LTE	L-Rx	−2.0	−4.4	−2.4	2.1
756	13/LTE	H-Rx	−1.8	−4.1	−2.1	2.0

the tunable matching circuit is lossy, the improvement in the mismatch outweighs the dissipative loss, thus improving the power transfer of the network and the efficiency of the entire antenna subsystem (antenna and matching circuitry). The initial data for this comparison used are the freespace performance. Table 8.2 shows the data for the LTE bands 12 and 13 that cover from 698 MHz, the lowest frequency of band 12, to 787 MHz, which is the highest frequency of band 13. In each case, the low and high channel results are shown for illustration. RTG as discussed in Equation 8.1 earlier, is the calculation of the improvement that the tuner provides over no matching network.

As the results show in Table 8.2, the loss due to mismatch has been considerably reduced with improvements of over 4 dB in some channels, and the overall benefit is close to 3 dB on average in the 700 MHz LTE bands. The return loss performance of the tuner under various settings is of some interest since it allows a designer to quickly see how well the antenna tuner is matching the input impedance. Figure 8.18 shows the return loss performance of the tuner under a number of different states across the entire frequency band of operation. Each return loss curve represents a different voltage setting on the PTICs. Since an analog voltage is used to bias each PTIC, optimum voltages are not constrained by capacitor steps and can be adjusted as desired to ensure the optimal performance from the network. Note that in Figure 8.18, only a small number of return loss curves are shown for illustration. This tuning capability leads to the easy adjustment of the PTIC values based on the handset use case. The use case can be determined either by triggering a proximity detector if the handset is held to the user's head or by using a more sophisticated method where mismatch detection is employed.[18] The tuner can then be adjusted to ensure that the mismatch is reduced to as low as possible.

In the case of head–hand loading, the mismatch loss as shown in Figure 8.15 is still over 4 dB at 700 MHz, which is slightly better than the freespace value. Since the mismatch loss is still high and also because there has been a frequency shift in the antenna performance due to the loading, the tunable matching network needs to be readjusted. This is carried out by changing the bias voltages and the resulting performance when compared with the no tuner condition is shown in Table 8.3.

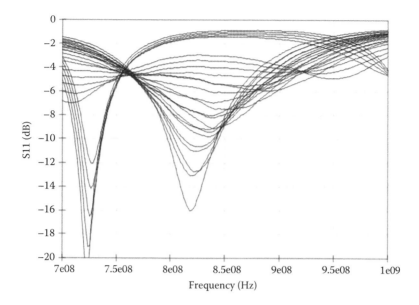

FIGURE 8.18   Return loss curves.

TABLE 8.3
Tuner Design Case Study Data in the 700 MHz LTE Bands for the BHHR Condition

Frequency (MHz)	Band/Mode	Channel	MML with Tuner (dB)	MML without Tuner (dB)	Tuner Insertion Loss (dB)	Relative Transducer Gain (dB)
698	12/LTE	L-Tx	−0.1	−4.2	−0.8	3.4
716	12/LTE	H-Tx	−0.9	−3.6	−1.5	2.1
728	12/LTE	L-Rx	−1.0	−3.2	−1.6	1.7
746	12/LTE	H-Rx	−1.0	−2.6	−1.3	1.3
777	13/LTE	L-Tx	−0.1	−1.6	−0.4	1.2
787	13/LTE	H-Tx	−0.1	−1.3	−0.4	0.8
746	13/LTE	L-Rx	−0.7	−2.6	−1.0	1.7
756	13/LTE	H-Rx	−0.4	−2.3	−0.7	1.6

The results in Table 8.3 show that the benefit of using the tuner is now close to 3.5 dB in some cases, and almost 2 dB on average in these 700 MHz LTE bands. This is a considerable increase in link margin and will improve handset performance in the field.

In the previous discussion, the RTG was computed using the performance with and without an antenna tuner. This is a common approach, but a more stringent and realistic comparison is to look at the performance of a fixed match under the same operating conditions. This will allow a full understanding of the benefit of the variable tuner over a fixed match under the two use cases that have been examined in this section. The design of a fixed matching network is a subjective exercise since a designer is faced with a trade-off of whether to design for a single use case, such as freespace; a more realistic condition, such as hand or head-and-hand loading; or something in between. In many cases, the simple freespace use case is used as it requires no special test equipment and can be done relatively quickly by hand using a VNA. Of course, this is a non-optimum approach and will lead to lower performance than could be obtained at some frequencies and use cases. In this example, the commercially available Optenni Lab simulator will be used along with the freespace S-parameters. A four-element network will be considered and the result of the automatic optimization process is shown in Figure 8.19.

The automatic optimization carried out using the commercially available simulator provides the user with the options to create all permutations of the four-element match, or simply output the best result based on the efficiency as calculated in each band. The results shown here are for the final selected circuit and the performance has been analyzed using the same Q values for the inductors as used in the tunable network and typical high Q capacitors values of around 200. The results for the 700 MHz LTE bands have been tabulated in the same way that they were in the tunable network case and are shown in Table 8.4 for freespace operation.

Comparing the results for freespace operation between the fixed matching circuit and the tunable circuit in the 700 MHz frequency range, a few conclusions can be drawn. Looking at each of the transmit and receive channels, the tunable matching circuit outperforms the fixed match in every transmit and receive channel. This is an expected result since the tuner can be adjusted on a band-by-band and even channel-by-channel basis. The fixed-component matching network is by definition unchanging. It is also worth noting that in Band 12, which also includes the important Band 17, which is a subset of Band 12 used in North America and other countries, the improvement that using the antenna tuner yields is up to 1.3 dB. This can be simply achieved by using PTICs and associated control components.

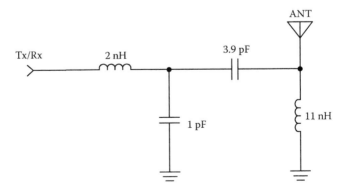

FIGURE 8.19   Fixed matching network for antenna example.

TABLE 8.4

Fixed Match Design Case Study Data in the 700 MHz LTE Bands for the Freespace Condition

Frequency (MHz)	Band/Mode	Channel	MML with Match (dB)	MML without Match (dB)	Match Insertion Loss (dB)	Relative Transducer Gain (dB)
698	12/LTE	L-Tx	−1.5	−5.8	−2.4	3.4
716	12/LTE	H-Tx	−1.5	−5.4	−2.3	3.1
728	12/LTE	L-Rx	−1.7	−5.1	−2.3	2.8
746	12/LTE	H-Rx	−1.7	−4.4	−2.2	2.2
777	13/LTE	L-Tx	−1.9	−3.3	−2.2	1.0
787	13/LTE	H-Tx	−2.3	−2.9	−2.1	0.9
746	13/LTE	L-Rx	−1.7	−4.4	−2.2	2.2
756	13/LTE	H-Rx	−1.9	−4.1	−2.3	1.8

TABLE 8.5

Fixed Match Design Case Study Data in the 700 MHz LTE Bands for the BHHR Use Case Condition

Frequency (MHz)	Band/Mode	Channel	MML with Match (dB)	MML without Match (dB)	Match Insertion Loss (dB)	Relative Transducer Gain (dB)
698	12/LTE	L-Tx	−0.9	−4.2	−1.5	2.7
716	12/LTE	H-Tx	−1.1	−3.6	−1.5	2.1
728	12/LTE	L-Rx	−1.4	−3.2	−1.7	1.5
746	12/LTE	H-Rx	−1.6	−2.6	−1.8	0.8
777	13/LTE	L-Tx	−1.9	−1.6	−2.0	−0.5
787	13/LTE	H-Tx	−2.2	−1.3	−2.0	−0.8
746	13/LTE	L-Rx	−1.6	−2.6	−1.8	0.8
756	13/LTE	H-Rx	−1.8	−2.3	−1.9	0.3

Under the BHHR use case, the same fixed matching network has been used, but the loaded S-parameters have been used instead of the freespace S-parameters for the efficiency and RTG calculations. Table 8.5 summarizes the results for the fixed match when operating under the BHHR use case condition and is shown later.

Looking at these results it can be seen that at some frequencies, the use of a fixed matching network actually increases the mismatch loss over a direct connection, and when compared with the tunable network, the antenna tuner provides up to 1.7 dB of improvement at some frequencies. The results show that tunable match provides better performance at every frequency considered here in the 700 MHz band, which is one of the most challenging bands to design for.

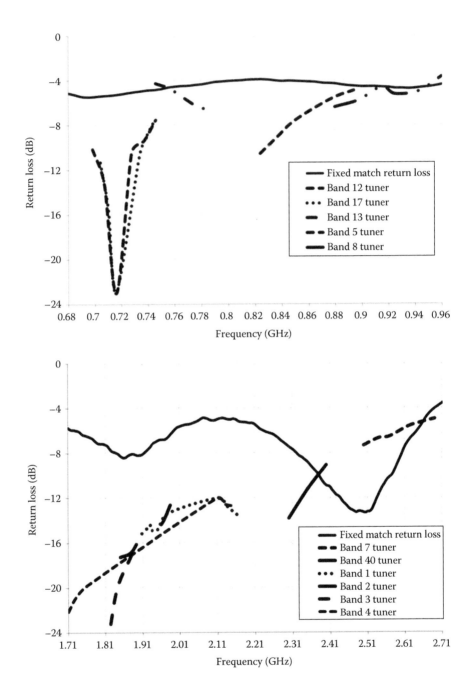

FIGURE 8.20   Return loss behavior for an antenna tuner and fixed matching network.

The results of the fixed match and the antenna tuner should also be compared fully across all of the frequencies of operation to get a full picture of the benefits and operation. This can be done most effectively by looking at the insertion loss of the fixed match and comparing it to the tunable match. It is also somewhat helpful to look at the return loss as well since this is a measure that many designers are more comfortable with, although this parameter would not be used to optimize the matching network for the reasons previously discussed. Figure 8.20 shows the return loss over frequency in both the low and mid/high bands for freespace operation. The antenna tuner results are shown in discrete bands since this is more applicable given that the bias voltages are adjusted for each band and set of subchannels. The data show that the tunable circuit outperforms the fixed matching network in virtually every band and channel and in some cases by quite a wide margin.

Figure 8.21 shows the companion charts for the insertion loss performance that show a similar trend with the antenna tuner outperforming the fixed circuit. However, in both the return loss and insertion loss charts, it can be seen that there are a very small number of channels where the fixed matching circuit offers slightly better performance than the tunable circuit. This is expected since it is natural to assume that the fixed matching circuit can be tuned to be optimal at some frequencies, however, at the vast majority of channels/bands, the antenna tuner offers superior performance. The results are similar for the BHHR use case and the insertion loss results for this case are shown in Figure 8.22. The tunable matching circuit outperforms the fixed circuit everywhere except for a few channels.

The performance of the antenna tuner will have a direct impact on the performance of the handset, and the improvement in the region of 2 dB shown here in the antenna subsystem will lead to a direct improvement in TRP, for example. This, in turn, leads to an improved link margin, which will lead to better coverage and higher data throughput. As shown in this example, there are very few channels where the fixed match can actually provide better performance than a tunable network, and even in those cases, the delta is low. The benefit of using PTICs in a tunable matching network over the traditional fixed-component approach is clear.

8.7 PTIC ANTENNA TUNER DEPLOYMENT

The implementation of a feedpoint antenna tuner is conceptually straightforward and many new handsets have been deployed using this technology. Figure 8.23 shows two examples where the PTICs are being used. One example shows the plastic packaged version of the PTICs and the other shows the chip scale package. Both are placed relatively close to the antenna and only occupy a small footprint. Also the inductors used to complete the full tuner circuit are shown. High-quality RF wirewound devices are recommended to ensure best performance.

The exact number of tuning devices and final schematic vary between handset manufacturers as well as where the tuner(s) is deployed in the handset, such as the main antenna, the diversity antenna, or both. Also, the use of closed- or openloop tuning solutions varies from handset to handset. The most common type of antenna tuner that is used today is the open-loop version, where the bias voltages

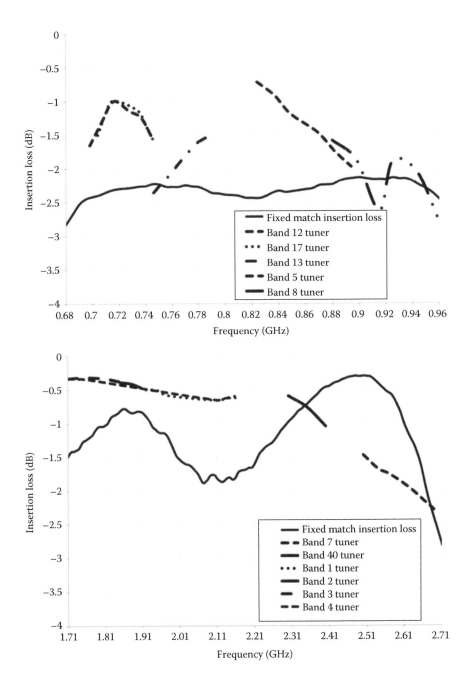

FIGURE 8.21   Insertion loss behavior for an antenna tuner and fixed matching network: Freespace.

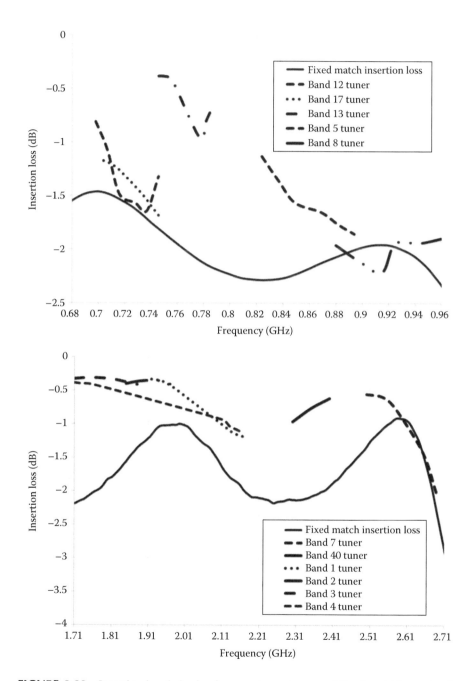

FIGURE 8.22 Insertion loss behavior for an antenna tuner and fixed matching network: BHHR.

FIGURE 8.23   Photographs of PTIC antenna tuners in commercially available devices.

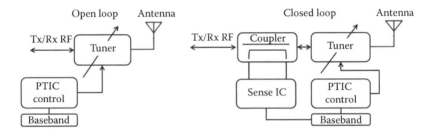

FIGURE 8.24   Block diagram comparing open- and closed-loop antenna tuner systems.

for each band/channel and use case are predetermined and stored in lookup tables, which are then used by the handset during operation on the relevant channel, and so on. Figure 8.24 shows simplified block diagrams for both open- and closed-loop antenna matching systems. Open-loop tuners are only able to adjust to the use cases that were preset when the design work was undertaken. In contrast, PTIC closed-loop antenna tuning systems use a feedback mechanism to continuously adjust the matching circuit to ensure the lowest possible insertion loss of the matching network. This results in the maximum power transfer and the optimal performance of the handset. Closed-loop antenna tuner systems have the big advantage that they can more easily adapt to the changing environment of the phone and antenna. The system uses a measurement IC that allows the impedance to be measured, which, when combined with the known S-parameters of the tuner, allows the optimum capacitance values to be selected irrespective of the use case that the phone is in. The design of the actual tunable matching network for a closed-loop matching system still follows the same methodology as outlined earlier. The difference between open and closed-loop operation lies in the additional components used to control the tuner. PTICs are ideally suited to operate in either open- or closed-loop implementations and are currently being used in both configurations in a number of commercially shipping handsets.

8.8 CONCLUSIONS

This chapter has reviewed how the wide frequency range of operation of the antenna(s) in a modern handset usually results in mismatch-limited performance. Typically, this mismatch occurs at the band edges and in the 700–960 low-frequency bands, but can also occur in the mid- and high-frequency bands. Additionally, the loading caused by use cases, such as head-and-hand or hand, can detune the antenna, making the design of a fixed match almost impossible to optimize over all frequencies and use cases, resulting in suboptimal performance. The use of PTICs to replace fixed capacitors results in a tunable network with considerable flexibility. The design of a tunable network can accommodate different use cases and should be optimized for insertion loss rather than return loss to ensure optimal performance. The tuner networks can utilize one, two, or three tunable elements, and the exact selection will depend on how strongly mismatched a given antenna is, the required bands and bandwidth needed, and the insertion loss that can be tolerated. The tunable network can be adjusted for each use case and used in either an open-loop or closed-loop antenna tuner systems. When compared with a fixed match, the tunable network will provide significant improvements in virtually all bands/channels and use cases, and this improvement will lead to improved performance in the field. The improvements that can be achieved using an antenna tuner with PTICs require very little additional circuitry beyond the tunable capacitors themselves and a central control IC. The use of a central control IC allows for a very flexible approach to antenna tuning and multiple antennas can be tuned using a single controller. Many handsets are already shipping with both open- and closed-loop PTIC antenna tuners, and the technology is now moving from the high-end smartphone segment into the mass-market segment showing that the adoption is now common and a standard handset feature.

REFERENCES

1. 3rd Generation Partnership Project Specifications, www.3gpp.org.
2. Optenni Lab Matching Network software, Optenni Ltd., Espoo, Finland, www.optenni. com.
3. Advanced Design System by Keysight Technologies, www.keysight.com.
4. Microwave Office by National Instruments AWR, www.ni.com/rf/awr/.
5. Chen, L., Forse, R., Chase, D., and York, R., Analog tunable matching network using integrated thin film BST capacitors, *IEEE MTT-S International Microwave Symposium Digest*, June 2004, Fort Worth, TX, pp. 261–624.
6. Dongjiang, Q., Yu, Z., Tsaipi, H., Kimball, D. et al., Antenna impedance mismatch measurements and correction for adaptive CDMA transceivers, *IEEE MTT-S International Microwave Symposium Digest*, June 2005, Atlanta, GA, pp. 783–786.
7. Ali, S. and Payandehjoo, K., Tunable antenna techniques for compact handset applications, *IET Microwaves, Antennas and Propagation*, 8(6), 401–408, April 2014.
8. Rhode and Schwarz Vector Network Analyzers, www.rhode-schwarz.com.
9. Keysight Technologies' Vector Network Analyzers, www.keysight.com.
10. Yip, P., *High Frequency Circuit Design and Measurements*, 1st edn., Chapman & Hall, London, UK, 1990.

11. Schmidt, M., Lourandakis, E., Leidl, A., Seitz, S., and Weigel, R., A comparison of tunable ferroelectric Pi and T-matching networks, *Proceedings of the 37th European Microwave Conference*, 2007, Munich, Germany, pp. 99–101.
12. Gu, Q. and Morris, A., A new method for matching network adaptive control, *IEEE Transactions on Microwave Theory and Techniques*, 61(1), 587–595, January 2013.
13. Collin, R.E., *Foundations for Microwave Engineering*, 2nd edn., McGraw-Hill, New York, 1992.
14. Rahola, J., Power waves and conjugate matching, *IEEE Transactions on Circuits and Systems—II: Express Briefs*, 55(1), 92–96, January 2008.
15. Rahola, J., Estimating the performance of matching circuits for antennas, *Proceedings of the Fifth European Conference on Antennas and Propagation*, Barcelona, Spain, 2010, pp. 1–3.
16. Rahola, J., Optimization of matching circuits for antennas, *Proceedings of Fifth European Conference on Antennas and Propagation*, 2011, Rome, Italy, pp. 776–778.
17. Tunable Capacitor Control IC Datasheet—TCC-106, www.onsemi.com.
18. Boyle, K.R., de Jongh, E., Sato, S., Bakker, T., and van Bezooijen, A., A self contained adaptive antenna tuner for mobile phones, *Sixth European Conference on Antennas and Propagation*, March 2012, Prague, Czech Republic, pp. 1804–1808.

9 Aperture-Tunable Antennas

Handset OEM Perspective

Ping Shi

CONTENTS

9.1 INTRODUCTION

Development of mobile communications has drastically changed the handset antenna design landscape. The handset antenna design target has changed from gain and radiation pattern to system efficiency and better user experience for multiband and multimode applications. The traditional approach in passive multiband antenna design is to trade antenna efficiency with bandwidth. Antenna tuning was proposed to address the multiband requirements. There are two ways to tune the antenna for multiple band operation: impedance tuning and aperture tuning. Antenna tuning has become practical in handsets with recent advancements in radio frequency (RF) tuning components: varactors, MEMS tunable capacitor, pin-diodes, RF switches, and so on.

The adoption of LTE-Advanced[1] presents a new set of requirements to tunable antenna technology. In this chapter, tunable antenna design challenges will be presented first, followed by a survey of tunable antenna technology. Tunable antenna design and verification methods will be discussed next. A summary of antenna tuning will conclude this chapter.

9.2 TUNABLE ANTENNA DESIGN CHALLENGES

Since the commercial introduction of LTE, a typical smartphone will support not only LTE but also legacy 2G/3G technologies such as GSM/EDGE, CDMA 1x_EVDO, WCDMA, and/or HSPA. In a state-of-the-art mobile phone, the main antenna is required to support from 699 to 960 MHz in low band and 1710 to 2170 MHz in the middle band along with 2300 to 2690 MHz in the high band. Multiple antennas are required to support receive diversity, Wi-Fi, GPS, and so on. With the introduction of LTE-Advanced, we are seeing more mode and band combinations from the standards as well as from regional market requirements. For example, carrier aggregation (CA) has been introduced to increase usable bandwidth by aggregating two or more carriers in the downlink and/or uplink. In some regional markets, dual sim, dual active (DSDA) functionality has been introduced to support simultaneous call states on two networks, for business reasons such as roaming. The transition from legacy 2G/3G networks also introduced some demands like simultaneous CDMA voice and LTE (SVLTE) and simultaneous GSM and LTE (SGLTE). Some of the demands like SVLTE/SGLTE will gradually phase out as network upgrades progress and spectrum refarming starts and others like DSDA will purely depend on business decisions. Here, we will focus on the emerging requirements from LTE-Advanced, which includes both downlink CA and uplink CA, and their impact on tunable antenna design.

The first challenge arising from CA is wider instantaneous bandwidth. 3GPP categorizes LTE interband CA into several types according to the band separation, harmonic relationship, and intermodulation product, as shown in Table 9.1.

A single main antenna architecture is used for standardization purposes as shown in Figures 9.1 and 9.2.

In this architecture, a diplexer is used to support simultaneous high–low operation (class A1/A2); a quadplexer is used to support simultaneous high–high or low–low operation (class A3/A4). The quadplexer could be constructed either from discrete

TABLE 9.1
LTE Interband CA Classes

Carrier Aggregation Class	Definition	Examples
A1	Low-band–high-band combination without harmonic relation between bands	B5_B2
A2	Low-band–high-band combination with harmonic relation between bands	B12_B4, B17_B4, B8_B3
A3	Low-band–low-band or high-band–high-band combination without intermodulation problem (low-order IM)	B12_B5, B1_B3
A4	Low-band–low-band or high-band–high-band combination with intermodulation problem (low-order IM)	B8_B20
A5	Combination except for A1–A4	

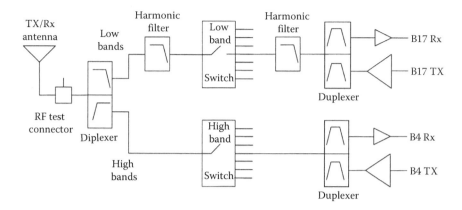

FIGURE 9.1 Single main antenna architecture v1.

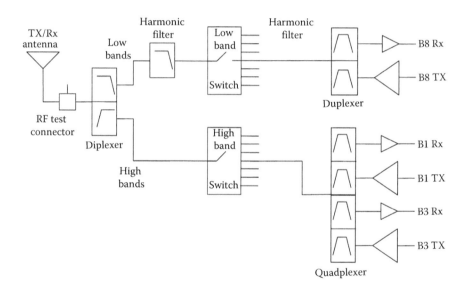

FIGURE 9.2 Single main antenna architecture v2.

duplexers or integrated on one die or within a module. For three-carrier interband CA (class A3/A4), a hexaplexer will be used. LTE CA requires wider instantaneous antenna bandwidth. The instantaneous bandwidth could be both in low band or both in high band (class A3, A4) or in low band and high band (class A1, A2). With support of multiple markets/carriers, this requirement presents a different challenge for antenna design, especially for tunable antenna design.

A second challenge arising from LTE-Advanced is the linearity requirement for the system. To support LTE CA band combination like class A2 and A4, a high linearity RF system design is implemented to minimize TX harmonic or intermodulation product. As shown in Figure 9.1, the reference RF front-end design minimizes

TX harmonic and intermodulation products by using high linearity components like duplexers and RF switches and applies filtering using high rejection diplexer or harmonic filters. A typical example for the harmonic relationships encountered in CA is the LTE CA band combination of B17_B4 with B17 as primary component carrier (PCC), as shown in Figure 9.3. The third harmonic of PCC TX (B17 TX) is generated from the PA, which is partially suppressed by duplexer, and from the duplexer, which is partially suppressed by following harmonic notch filter, and finally from antenna switch module (ASM), which in turn is suppressed by the diplexer.

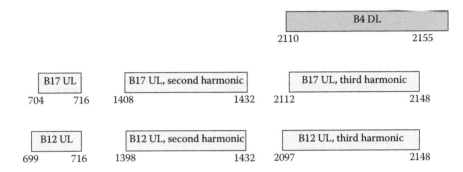

FIGURE 9.3 Example of LTE CA band harmonic relationship: B17_B4.

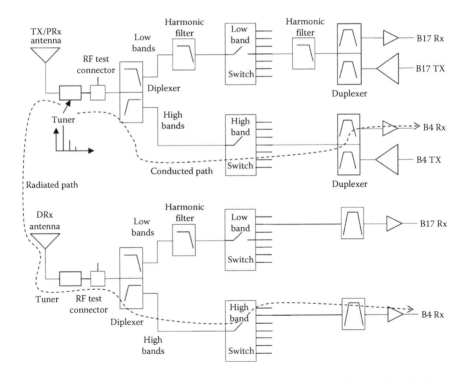

FIGURE 9.4 Coupling of TX harmonic or intermodulation product into receiver chain.

For handset antenna design, traditional passive antenna designs work well due to their passive and high linearity nature. For a tunable antenna design, tuning components are placed close to or on the antenna. The generated TX harmonic or intermodulation product will couple into the receiver chain through two different paths: conducted path to the PRx (Primary R) chain and radiated path to DRx (Diversity R), as shown in Figure 9.4. For both paths, filtering cannot be used to suppress the harmonic or intermodulation interference falling into RX band. The generated harmonic or intermodulation product has to be small enough to cause no desense to either the primary or the diversity receiver. This requirement presents a serious challenge in the design of a tunable antenna solution.

9.3 SURVEY OF TUNABLE ANTENNA TECHNOLOGY

Tunable impedance matching methods optimize system efficiency by reducing the return loss and increasing transducer gain for a specific frequency band. This type of matching approach fits the global branding strategy used by handset original equipment manufacturers (OEMs) very well, as it can quickly change an antenna optimized for one set of frequency bands to support another set of frequency bands without the need for industrial design, mechanical design, or antenna design changes. Massive flagship smartphones implementing tunable impedance matching techniques have been shipped by major handset OEMs recently, reflecting the growing tremendous success of the impedance tuning technique. Unlike the antenna impedance matching network, aperture tuning reconfigures the antenna to a different length or shape electrically, thus changing the antenna aperture. Aperture tuning can achieve performance (improvements) that cannot be rivaled by impedance tuning. By reusing the entire antenna volume for different operating frequency band, the physical dimensions of the antenna can be reduced. Providing narrower instantaneous bands that can be tuned, rather than covering all the bands simultaneously, interference can also be reduced.

9.3.1 TUNABLE IMPEDANCE MATCHING

Tunable matching techniques improve antenna efficiency by reducing the mismatch loss and increasing transducer gain. Considerable work has been done in universal tunable matching network design[2,3] and adaptive impedance tuning techniques.[4–6] The deployment of LTE-Advanced requires wider simultaneous antenna bandwidth and flexible band combinations, which presents new challenges for antenna impedance techniques.

LTE CA with high-band–low-band combination and high-band–high-band combination are widely seen as being in the early deployment phase. Wideband matching techniques[7] could be used to provide simultaneous matching of both low band and high band, but the difficulty of providing such solutions increases quickly as more band combinations need to be supported. Additional tunable components are required and insertion loss increases due to the added tuning components. To relax the tuner requirements, antenna could be designed to provide wider bandwidth covering entire mid/high band with good return loss. Impedance tuning then is only applied to low band, thus reducing the required tuning component count. With this consideration, the tunable matching network is designed to be a high-pass filter with near 50 ohm impedances at high/mid band and with an optimum matching topology for the low-band antenna impedance. For example, instead of using an individual L or C as in a typical matching network, an LC tank, serial LC circuit or parallel LC circuit is used to provide the desired frequency response (tank resonance frequency is carefully chosen to have a high-pass response while effectively matching the low-band), as shown in Figure 9.5. Band-pass fitter design techniques could be used in matching network design.

Antennas with this property draw considerable attention in practical product design. These antennas usually have very wide radiation bandwidth, with different radiation elements for low band and high band, and very little coupling between low band and high band. For example, a T-shape monopole antenna typically radiated in high band using two radiation arms and using the board printed circuit board (PCB) ground at low band. This type of antenna has a very wide radiation bandwidth, making it a good candidate for handset application. Other similar antennas include segmented inverted F antenna (IFA) antennas and Pi-shaped IFA antennas as shown in Figure 9.6.

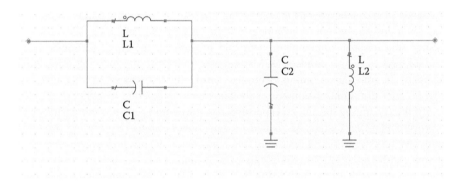

FIGURE 9.5   Frequency response using an LC tank circuit.

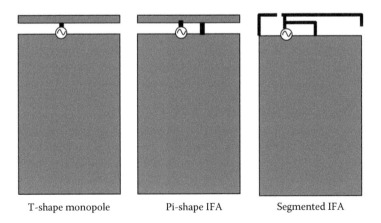

T-shape monopole   Pi-shape IFA   Segmented IFA

FIGURE 9.6   Examples of antenna structures with wide bandwidths.

9.3.2 Aperture-Tunable Antenna

Tunable impedance matching does not change antenna aperture; the system efficiency of the antenna with tunable impedance matching will not exceed the overall radiation efficiency. For a handset antenna design, the antenna is much shorter than the wavelength at low-band frequencies. The radiation efficiency drops quickly as it approaches the small antenna limit. This is where tunable matching techniques start to be less useful. On the other side, with aperture tuning, one can change the antenna radiation aperture and reshape the BW efficiency curve. With aperture tuning, antenna could be designed to have a narrow bandwidth with high efficiency while keeping the same volume, or to keep the same efficiency with reduced antenna volume.

With the wide variety of tuning components now available, aperture tuning methods are being aggressively explored for popular types of antenna designs. Planar inverted F antenna (PIFA) antennas were popular when PCB reuse between circuit and antenna was desired and the thickness of phone was not a concern. Aperture tuning of a PIFA antenna includes tuning the radiation arm length using pin-diodes[8] or loading PIFA antennas at different locations using tunable capacitors.[9] Monopole or IFA antennas are popular antennas for their high efficiency and broad bandwidth. Aperture tuning of monopole or IFA antennas includes extending/shorting the antenna with pin-diodes or tunable capacitors, loading the antenna with capacitance,[10] changing the shorting arms' length or shorting location.

By decoupling the high-band radiating element from the low-band radiating element and aperture tuning the low-band element, the aperture-tunable antenna technique can be applied to support LTE CA with a high-band–low-band combination (A1/A2 classes), and high–high combination (A3 class). Pi-shape IFA antennas and segmented IFA antenna are among the good candidates for this type of application. To support LTE CA with low-band–low-band combination, the antenna needs to support dual low-band resonance. Metamaterial inspired antennas are good candidates.[11] This type of antenna provides two low-band resonances: one from the normal radiation mode (L–C) and another from the left-hand modes (C–L), as illustrated in Figure 9.7.

Aperture tuning is applied to tune the left-hand resonance mode at the coupling (equivalent C1) or the grounding (L2). Similar designs have been used by several handset manufactures.

9.4 DESIGN AND VERIFICATION METHODS

Application of tunable antenna designs to handsets is determined by several factors, including, but not limited to, bill-of-materials (BOM) cost, industry design (ID), supported bands and technologies (BAT). The ID will determine what types of antenna can be used. BOM cost will determine whether a tunable solution is affordable and if so, what type of tuning components and approach will be used. The supported bands and technologies will affect the antenna-type selection and tuning technology, and how the trade-off should be made between antenna performance, industry design, and BOM cost.

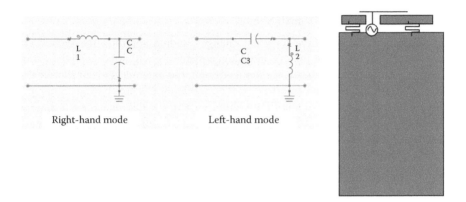

FIGURE 9.7   Metamaterial-inspired dual low-band resonance antenna.

It is desirable to have a passive wideband antenna to support all LTE bands and CA combinations. However, there are cases where antenna design is limited by ID, for example, metal ID such as metal ring, metal back IDs. In these cases, a tunable antenna is typically used to provide the same functionality with less volume. To meet requirements from LTE CA, a tunable antenna has to support much wider instantaneous bandwidth, thus imposing new limitation on the antenna design and performance.

9.4.1 SELECTION OF ANTENNA AND TUNING COMPONENTS

The handset antenna design starts with antenna type selection. Handset architecture determines a few usable antenna types, as do the outcome of ID design and internal mechanical design, components placement, and PCB layout. Once antenna type is determined and antenna tuning is determined to be required, a candidate tuning solution is selected from among the many options. Tunable components are selected according to the tuning method to be employed. Two major types of tuning components are available for antenna tuning today: tunable capacitors and RF switches (with fixed capacitors).

Tunable capacitors provide continuous capacitance within a range. Tunable capacitors can be realized by varactor diodes, barium strontium titanate (BST) film capacitors, switched GaAs/CMOS capacitor arrays, or radio frequency microelectromechanical system (RF-MEMS) capacitor arrays. The tuning mechanism and control interface also determine the application of tunable components. The capacitance of varactor diode and BST film capacitor is controlled by the voltage across the capacitors. Varactor diodes and BST film capacitors do not have a reference ground. These two types of tunable capacitors are applied where the parasitic capacitance to the ground is critical. In some aperture-tunable designs, capacitance between different radiators is tuned to change the antenna aperture. MEMS and switched CMOS/GaAs capacitor arrays change the capacitance by switching on and off component capacitors. Capacitance values are digitized to discrete values covering the tuning range. These two types of tunable capacitors usually have a digital control interface like RF-MIPI or SPI/I2C. For MEMS tunable capacitors, internal charge pumps are implemented to drive each element capacitor to high or low-capacitance values. These two types of tunable capacitors have some parasitic to ground and require ground/power/digital interface to control the capacitance values. These requirements could limit their application in certain aperture-tunable antenna designs where ground/power/digital interface is not available or parasitic to ground should be extremely low. Typically, digital tunable capacitor provides a minimum of 16 states with a tuning range about 4:1 or higher. A higher tuning range with lower parasitic capacitance is preferred as it enables "universal matching network" and provides flexibility in aperture tuning.

Radio frequency switches make inductance tunable. There are two topologies for RF switch: serial switch or serial shunt switch. Serial shunt RF switches can provide extra isolation between ports at the cost of shunting another port to ground. The serial configuration switch is the most popular RF switch topology for antenna tuning. RF switches are widely used in aperture tuning, where switches can be used to extend or shorten radiator length, to create a shorting arm to ground, or to change the grounding inductance.

The capacitance tuning range (and tuning ratio) and Q (Quality factor) are among the most important electrical parameters for tunable capacitors. RF switch structure, on-resistance and off-capacitance determine how the switch is used for antenna tuning. To have a high linearity antenna, higher IIP2, IIP3, voltage handling capability, and power handling capability are essential for all tunable components.

Special attention should be paid to power handling and voltage handling capability during component selection. Power handling capability is usually specified in a 50 ohm environment. However, the switches or capacitors used in impedance matching or aperture tuning are not in a 50 ohm environment. Voltage handling capability is more useful in a non-50 ohm environment. Different component suppliers specify voltage handling capability differently: some of them use rms voltage, while others use peak voltage. This voltage is also dependent on the technology, for example, complementary metal oxide semiconductor (CMOS) and pseudomorphic high electron mobility transistor (PHMET) will have different breakdown mechanism; while MEMS capacitors have the pull-in voltage and pull-out voltage considerations. These differences should be considered during the tuning component selection process in order to decide between different technologies and different vendors.

9.4.2 TUNABLE ANTENNA DESIGN

For an impedance tunable antenna, the antenna design goal is to have radiation efficiency as high as possible for the applicable bands. The raw antenna return loss is expected to be better than a threshold value. Because of the limited tuning range of a tunable capacitor, a tuning network can only tune an antenna well within a range. Due to this limited capacitance range, the tuning ratio (max/min ratio) of the capacitors, and achievable values of Q, there is no universal matching network with acceptable insertion loss in the near future. The tunable matching network should be optimized for the raw antenna impedance. The antenna and matching network design is an iterative process.

For aperture-tunable antennas, the tuning component must be included in the initial design and simulation. It will be desirable to have the tuning component or circuit in the initial mock-up antenna. Often the tuning components or circuits are not available when needed for one reason or another. It is a common practice to use a passive component to simulate the tuning component in the mock-up design with RLC models provided by the vendors. The same technique is applied for antenna simulation, where a simple RLC model of device is good enough for the initial design. In the initial design, the return loss and antenna efficiency are among the major design targets. Next, antenna performance over different use cases such as beside head, beside head and hand, various hand positions should be considered to validate the antenna performance. Specific absorption rate/hearing aid compatibility (SAR/HAC) should also be evaluated during this phase. Any failure to achieve required performance specifications for the earlier items will trigger another iteration of the design.

For tunable antenna design, antenna linearity must be evaluated before antenna design is locked down. Traditionally, the linearity concern is focused on spurious emissions. Residual spurious emissions are specified by individual government agencies such as FCC(U.S.)/Industry Canada or EC, as well as by standard organizations like the European Telecommunications Standards Institute (ETSI) and Cellular Telecommunication Industry Association (CTIA). Compared to the radiated spurious emissions (RSE) requirements, the harmonic desense to receiving bands in LTE CA is more challenging. In a 50-ohm system, this is equivalent to have the impedance tuner with IIP3 over 81 dBm. When the system impedance is not 50 ohm, the voltage across the component will be higher and correspondingly the requirement to IIP3 will be higher. It is necessary to simulate or measure actual generated harmonic or intermodulation to validate antenna design.

Antenna simulation could be performed in a frequency domain and time domain. Typically, simulation methods could be FDTD, FEM, or a hybrid. All of the earlier mentioned methods terminate simulation based on the convergence of simulation. For example, if the difference between two simulations is less than a threshold, the program will determine that the simulation reaches the convergence and will terminate the simulation. The actual termination criteria varies, with typically −30 to −40 dBc. A typical IIP3 of tuning components is much better and the generated harmonic is far below −40 dBc. Any nonlinear contribution generated from the tuning component cannot be simulated by antenna simulator. A joint simulation between antenna simulator and circuit simulator is required to obtain the harmonic or intermodulation product level.

9.4.3 SIMULATION OF HARMONICS AND INTERMODULATION FROM TUNABLE ANTENNAS

Nonlinearity from a tunable antenna impacts radio performance in at least three different ways. It can (1) cause interference to a radio in close proximity, (2) desense the primary receiver operating at a different band, or (3) desense a diversity receiver operating at a different band of the same handset.

RSE is used to characterize spurious emission interference to radios in close proximities. In the United States, the RF bands used by the mobile communication industry are named as Cellular/PCS/AWS/IMS band. Different FCC rules are applied for each individual band as specified in FCC parts: 15C, 22, 24, 90, and so on. Emission requirements are different between countries. Standard organizations like ETSI and CTIA created RSE specification to meet most stringent requirement across countries. The limit for RSE is usually specified in effective radiated power (ERP), with the limit of −36 dBm using 100 kHz resolution bandwidth in the range of 30–1000 MHz and −30 dBm using 1 MHz resolution bandwidth in range of 1000 MHz to 12.75 GHz. For a typical multimode phone supporting GSM/CDMA/WCDMA/LTE, the most challenging case is RSE for low-band GSM at 33 dBm power.

Simulation is used to evaluate antenna nonlinearity without building a PCB board and mocking up an antenna. Simulating the nonlinearity would require a large signal model for the tuning components and an *S*-parameter model for antenna. The large signal model could be a physical model, an empirical model, or measured X-parameters. For antenna designs using an impedance tuner, the antenna is represented as a one-port device when the simulation is used to evaluate RSE or harmonic/intermodulation generated from the antenna to the main receiver. When the goal is to evaluate the harmonic/intermodulation desense to diversity receivers, the antenna system, including both main and diversity antenna, is used and represented as a two-port device: the feed to main/primary antenna is one port, the feed/input to secondary/diversity antenna is the second port. A one-port or two-port *S*-parameter model will be generated from antenna simulators such as XFDTD, CST, SEMCAD, or HFSS.

In aperture-tunable antennas, tunable components are imbedded into the physical antenna design. As antenna simulators generally do not simulate nonlinearity (well), nonlinear components must be de-embedded from antenna port. The tuning component is modeled as an RCL component; a port is defined at each tuning component, along with a port for the main antenna feed (and diversity antenna feed). An n-port *S*-parameter will be generated from antenna simulation. To evaluate RSE and the impact to the main receiver, an aperture-tunable antenna with n-tunable components is represented by an (n + 1)-port device: antenna feed is one port, the rest of n ports are defined at n tuning components. To evaluate desense to the diversity receiver, the antenna system (with an aperture-tunable antenna using n tuning component) is represented by (n + 2)-port devices: the main antenna feed is one port, the diversity antenna feed is one port, and the rest ports are defined at n tuning components. Figure 9.8 shows a harmonic balance simulation schematic, where the antenna is modeled as a two-port data component with one port as feed and tuning component at another port. The nonlinear model of tuning component is from the measured X-parameter. In Figure 9.9, an antenna system (main/diversity) and one tuning component are modeled as a three-port device; the tuning component is represented using a measured X-parameter.

Circuit simulators could be any simulator such as Ansys DesignerRF or Keysight Advanced Design System (ADS) as long as it supports RF *S*-parameter simulation and harmonic balance simulation. Some RF simulators can (directly) communicate with antenna simulators and perform joint (concurrent) circuit/antenna simulation.

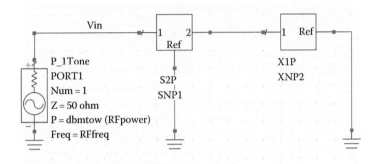

FIGURE 9.8   Schematic for harmonic balance simulation.

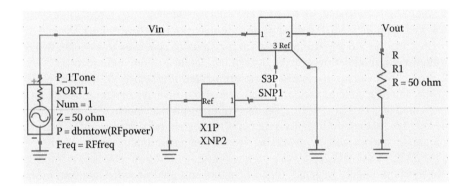

FIGURE 9.9   Model of a main/diversity antenna system with one tuning component.

9.4.4   MEASUREMENT OF HARMONICS AND INTERMODULATION FROM TUNABLE ANTENNAS

Harmonic and intermodulation resulting from the implementation of a tunable antenna should be measured once the antenna is available for characterization.

Residue spurious emission measurement is well defined. The measurement is performed in an anechoic environment where the reflection from the environment is negligible. At a predefined distance, a spectrum analyzer is connected to a wideband receiving antenna. At a transmit frequency, the handset device is rotated to find the directional angle for the maximum antenna gain. At that maximal gain angle, a spectrum analyzer is used to sweep the frequency of interest and obtain the residue spurious emission. The spurious emission power is calibrated using the ratio of spurious emission power to total TX power. Antenna radiation pattern differences between TX frequency band and the spurious emission frequency band should be considered when simulation is used to evaluate the RSE.

The generated harmonic or intermodulation product can be measured as shown in Figures 9.10 and 9.11. First, the RF signal is generated from the signal generator and amplified by a high-power amplifier. Isolators are used to eliminate load impact

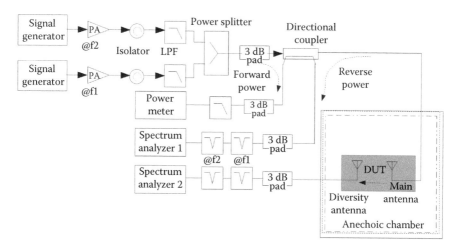

FIGURE 9.10   Generated intermodulation product measurement system.

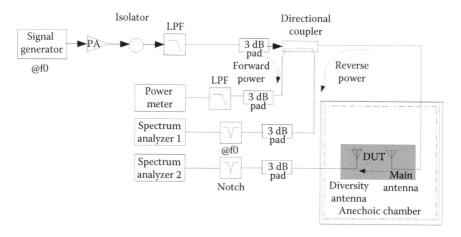

FIGURE 9.11   A generated harmonic power measurement system.

to the power amplifier. Depending on whether intermodulation or harmonic is measured, Figure 9.10 or Figure 9.11 will be used.

The signal (or combined signal) is then passed through a directional coupler, from which both forwarded power and reversed power are monitored. In the forward path, another low-pass filter (or notch at harmonic frequency) is used to remove the harmonic generated from the power detector from the RF system. In the reverse path from the directional coupler, one- or two-notch filters are used to remove the fundamental power that could limit the measurement accuracy because of the limited dynamic range of the spectrum analyzer and limited linearity of the system.

The antenna to be measured should be placed in an anechoic environment where the reflected power is negligible. The measured harmonic or intermodulation from reverse path by spectrum analyzer #1 is the interference that will fall into the main/primary receiver band. To measure the intermodulation product and harmonic received by diversity RX chain, a second spectrum analyzer is connected to diversity antenna port as shown in Figures 9.10 and 9.11. The measured harmonic or inter-modulation from diversity antenna port by spectrum analyzer #2 is the interference that will fall into the diversity receiver RX band.

9.5   CONCLUSIONS

This chapter addresses two important aspects of a tunable antenna design simultaneous bandwidth and linearity. With the rapidly growing commercial success of the tuning component industry, it is expected that tuning components with extraordinary high IIP3/IIP2 and Q will emerge in the near future. High tuning ratio, low parasitic tunable capacitor technology will enable true universal matching network and high(er) order matching networks. Closed-loop tuning solutions based on these types of capacitors will capture and increase the amount of market share. Aperture-tunable methods will be explored for more antenna types. It is also expected that aperture-tunable antenna designs will demonstrate their value in the vast majority of future handset products.

REFERENCES

1. 3GPP, http://www.3gpp.org.
2. Sun, Y. and Fidler, J.K., Practical considerations of impedance matching network, *HF Radio Systems and Techniques*, 1994, York, UK, pp. 229–233.
3. Sun, Y. and Fidler, J.K., Design method for impedance matching networks, *IEE Proceedings on Circuits, Devices and Systems*, 1996, pp. 186–194.
4. Boyle, K., Bakker, T., de Jongh, M., and van Bezooijen, A., Real-time adaptation of mobile antenna impedance matching, *Antennas and Propagation Conference (LAPC)*, 2010, Loughborough, UK, pp. 22–25.
5. van Bezooijen, A., de Jongh, M.A., van Straten, F., Mahmoudi, R., and van Roermund, A., Adaptive impedance-matching techniques for controlling L networks, *IEEE Transactions on Circuits and Systems*, 57(2), February 2010, pp. 495–505.
6. Gu, Q. and Morris, III, A.S., A new method for matching network adaptive control, *IEEE Transactions on MTT*, 61(1), January 2013, pp. 587–595.
7. Yarman, B.S., *Design of Ultra Wideband Antenna Matching Networks: Via Simplified Real Frequency Technique*, Springer, 2008.
8. Komulainen, M., Berg, M., Jantunen, H., Salonen, E.T., and Free, C., A frequency tuning method for a planar inverted-F antenna, *IEEE Transactions on Antenna and Propagation*, 56(4), April 2008, pp. 944–950.
9. Panayi, P.K., Al-Nuaimi, M., and Ivrissimtzis, L.P., Tuning techniques for the planar inverted-F antenna, *IEE National Conference on Antennas and Propagation*, 1999, York, UK, pp. 259–262.
10. Tornatta, P.A. and Gaddi, R., Aperture tuned antennas for 3G-4G applications using MEMS digital variable capacitor, *IEEE MTT-S International Microwave Symposium Digest (IMS)*, 2013, Seattle, WA, pp. 1–4.
11. Lopez, N., Lee, C.J., Gummalla, A., and Achour, M., Compact metamaterial antenna array for long term evolution (LTE) handset application, *IEEE International Workshop on Antenna Technology, iWAT 2009*, 2009, Santa Monica, CA, pp. 1–4.

10 Power Amplifier Envelope Tracking

Jeremy Hendy and Gerard Wimpenny

CONTENTS

10.1   WHAT IS ENVELOPE TRACKING?

Envelope tracking —ET for short—is a hardware technology that improves the energy efficiency of the RF power amplifier (PA) by dynamically modulating the power supply to the PA as the instantaneous amplitude ("envelope") of the RF signal changes.

ET can, in principle, be used in any high data-rate wireless transmitter, but the first market segment to adopt ET in significant volume has been the LTE smartphone market. Most high-end smartphones launched since mid-2014 use ET to improve battery life and reduce heat dissipation.

High data-rate wireless communications standards, like 4G/LTE and 802.11ac Wi-Fi, use signals which have a very high variation in magnitude. This increases the peak power needed to transmit the signal, although these peaks only happen relatively infrequently. The ratio of the peak power to the average power is known as the "peak to average power ratio" (PAPR).

In a conventional (non-ET) transmitter, the PA is supplied with a fixed DC supply voltage, which must be high enough to support the peak transmit power of the signal. The PA device itself is only energy efficient at the very peaks of the signal, where it approaches saturation. Consequently for much of the time, the supply voltage is higher than needed, causing power to be wasted in the form of heat dissipation. This is illustrated in Figure 10.1 by the rectangular gray area shown in the upper diagram.

In ET, the PA supply voltage is dynamically adapted in response to the instantaneous amplitude of the signal. This mode is shown in the lower half of Figure 10.1. By providing the PA with just enough supply voltage to operate, the energy efficiency of the PA is maximized at all times, not just at the peaks.

By keeping the PA in compression over much of the modulation cycle, the mean efficiency of the PA can be significantly increased—for an LTE uplink signal the increase is from around 30% with a fixed supply to around 55% with ET. When factoring in the power conversion efficiency of the ET power supply (around 80%), a net ET efficiency of around 45% can be achieved.

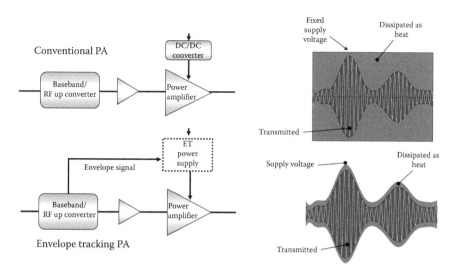

FIGURE 10.1 Overview of envelope tracking.

In addition to the energy saved, there is a significant reduction in thermal dissipation within the PA, which can help with both thermal design and RF performance.

Figure 10.2 illustrates the instantaneous efficiency curves of an ET PA, showing how the efficiency trajectory is maximized throughout the modulation cycle, by reducing the supply voltage as the output power reduces from the peak level,

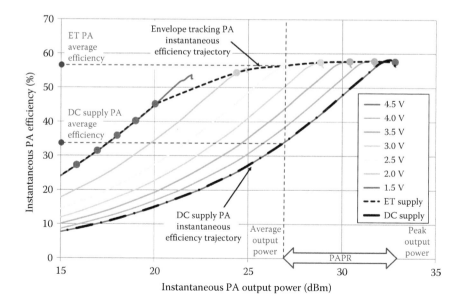

FIGURE 10.2 PA efficiency curves—envelope tracking versus fixed supply.

significantly increasing the average PA efficiency when used with a high PAPR waveform.

To implement, ET requires the following hardware components:

- A high-bandwidth envelope reference signal, generated by the modem or RFIC
- A high-bandwidth ET power supply chip (the "ETIC")
- An ET-capable or ET-optimized PA

10.2   GENERATING THE ENVELOPE REFERENCE SIGNAL

For an ET system to operate, the ET power supply must be provided with an envelope reference signal, which dynamically tracks the instantaneous amplitude (envelope) of the RF signal being transmitted.

In LTE handset applications, the envelope reference signal is normally a differential voltage generated by a dedicated high-speed DAC in the modem or RFIC. Interface drive levels and electrical characteristics of this interface have been standardized by the MIPI Alliance in the MIPI eTrak specification, originally released in 2013 and updated in late 2014—not to be confused with the MIPI RFFE control interface, which is also used for command and control of the ETIC.

For high-power infrastructure ET applications, parallel or serial digital interfaces (e.g., JESD204) are more common, with the ET DAC integrated into the ET power supply function, making it more straightforward to interface to programmable baseband SoCs or FPGAs.

A relatively simple digital signal processing chain is used to generate the envelope reference signal, usually from the digital I/Q signals, as shown in Figure 10.3.

Although straightforward, it requires a relatively high sample rate—typically 6× the RF channel bandwidth, that is, >120 MSPS for a 20 MHz LTE carrier, although some intermediate calculations may require higher sample rates.

The relationship between the instantaneous RF amplitude at the PA input and the supply voltage at the supply terminal of the PA must be precisely controlled. This is normally implemented in the form of a "shaping table"—a lookup table (LUT) in the baseband, which defines this nonlinear mapping.

The relative timing of the signals must also be controlled with subsample precision. The envelope reference signal must be generated slightly in advance of the RF signal at the PA input, to compensate for delays through the ETIC, and ensure precise timing alignment of the ET and RF inputs at the PA.

10.2.1   ENVELOPE PATH SIGNAL PROCESSING

The signal processing in the envelope path normally consists of the following steps:

- Coarse step (FIFO) programmable delay, usually in RF path
- Fine step (fractional delay filter) programmable delay, usually in ET path
- Upsampling/interpolation to 4×/6×/8× IQ sample rate

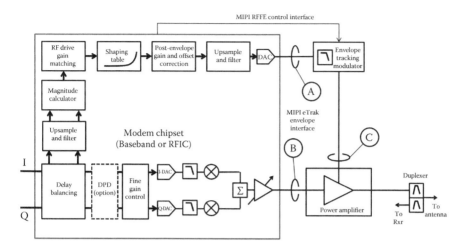

FIGURE 10.3   Envelope path generation logic.

- Magnitude calculation (CORDIC)
- Digital gain adjustment, for RF drive matching
- ET shaping table (LUT)
- Digital gain/offset adjustment, for DAC/ETIC analog errors
- Pre-DAC interpolation
- ET DAC
- Analog low-pass filter (normally implemented in the ETIC)

Depending on the architecture of the modem chipset, the envelope path logic may be implemented in either the baseband IC or RFIC. Architectures with a digital BB/RF interface will usually implement the ET function in the RFIC, and architectures with an analog BB/RF interface will usually implement the ET function in the BBIC, that is, the ET logic and DAC are in the same chip as the TX I/Q DACs.

The envelope processing function is relatively small, occupying 50k–100k gates of digital logic and consuming around 10 mW in a typical digital process.

10.2.2   Timing Alignment

Delay balancing, or timing alignment, between the ET and RF paths, is normally the first function to be performed, since this step is most economically performed at the lowest sample rate.

The requirements for timing alignment accuracy are directly related to the channel bandwidth; for 20 MHz LTE signals, an accuracy of ±0.5 ns is typically specified. This requires subsample timing resolution in the digital path.

It is important to note that timing alignment can be performed in either the ET or RF paths, since it is only the relative timing of the ET or RF paths at the PA (points "B" and "C" in Figure 10.3) that is of concern.

Most ET path implementations divide the timing alignment into coarse- and fine-grain steps. Coarse-grain timing can be implemented using a simple FIFO on the

IQ or ET path data, whereas fine-grain timing will require a fractional delay filter (FDF)—usually implemented on the ET path. The OpenET Alliance reference implementation uses a four-tap farrow filter (FIR) for 20 MHz LTE applications.

Although the eTrak envelope reference signal at point "A" must be generated slightly in advance of the RF signal at point "B," it is not always the case that the delay is added to the IQ path. If the RF path delay from IQ to PA input exceeds the ET path delay, the coarse-grain delay FIFO must be implemented into the ET path, rather than the IQ path.

The designer must fully consider all potential delays due to digital signal processing pipelining and analog filters in both ET and RF paths, which may also vary depending on channel bandwidth settings if there are switchable filters in the RF or ET paths.

The analog ETIC propagation delay will be specific to the ETIC design and may perhaps also vary with channel bandwidth setting if the ETIC includes switchable filters. As a guideline, around 10 ns of delay is typical for a handset ETIC.

The ET/RF timing alignment may require some form of temperature compensation, as the group delay of the ET and RF path filters is likely to vary by more than 0.5 ns over temperature extremes. Production-line calibration of timing alignment is also likely to be required, since the system timing alignment will be a function of both the RFIC and ETIC, as well as potential variations in the PA itself.

10.2.3   UPSAMPLING

Prior to calculating the magnitude of the IQ signal, the signal must be upsampled and filtered to provide sufficient bandwidth for the magnitude calculation, which is an often overlooked "pinch point" in terms of the sample rate. LTE IQ signals are normally generated at 1.536× the channel bandwidth (e.g., 30.72 MSPS for a 20 MHz channel). The magnitude calculation of the unshaped envelope must be performed at a higher sample rate to avoid aliasing, requiring interpolation of at least 4× to avoid unwanted distortion—6× (184.32 MSPS) or 8× (245.76 MSPS) are better choices.

10.2.4   MAGNITUDE CALCULATION

Calculating the magnitude of the IQ signal (i.e., the length of the vector specified by the IQ coordinates) is normally performed by a CORDIC function, which approximates the Cartesian-to-polar calculation without needing any hardware multipliers. CORDIC is an iterative algorithm, which in practice can be implemented as a hardware pipeline of n stages. Although the rule-of-thumb is that one CORDIC iteration is required for each equivalent bit of precision, simulations with LTE signals have shown that only eight iterations are required before the CORDIC becomes an insignificant noise source.

10.2.5   PRE-LUT GAIN ADJUSTMENT

Following the magnitude calculation, but before the shaping table calculation, a digital gain correction is performed. This is primarily used to compensate for variable gains in the analog RF path, for example, where a variable gain amplifier is used to attenuate the RF drive level to the PA input as the average power control level is changed.

This is illustrated in Figures 10.4 and 10.5, which show the envelope waveform as the average power level is backed off from 0 dBm at the PA input to −9 dBm at the PA input, for example by adding a 9 dB analog attenuation step into the transceiver RF path. It can be seen that simply scaling the ET output waveform after the shaping table would not give the correct supply voltage—the magnitude signal must be scaled before the shaping table, to ensure that the ET supply voltage remains correctly aligned with the RF drive level at the PA input.

This gain adjustment mechanism can also be used at production calibration to compensate for analog gain errors in the RF path; if closed-loop DPD is not employed, the RF and ET path gains must be matched to within ±0.25 dB for 20 MHz LTE.

10.2.6   ET Shaping Table

The ET shaping table is one of the most important control mechanisms in an ET system. It defines the relationship between the instantaneous envelope of the RF input to the PA and the instantaneous ET supply voltage and is usually optimized for the type of PA being used. The X axis of the shaping table represents RF signal level at the PA input (in digital terms, the gain-corrected IQ vector magnitude) and the Y axis represents the required ET supply voltage at the PA (in digital terms, the target ET DAC value).

Figures 10.4 and 10.5 illustrate a shaping table (top left of chart) and show how the PA input drive level (bottom left) maps to PA supply voltage (top right).

At a hardware level, the shaping table is usually implemented as a programmable LUT. The number of entries in the LUT and the interpolation method used between entries influence the PA linearity. For 20 MHz TD-LTE, 128 entries with linear interpolation provide sufficient accuracy, although it is possible to reduce the number of entries by providing a more complex hardware interpolation.

Although the shaping table may be characterized at design time in terms of PA input power (in dBm) and PA supply voltage (in volts), an embedded shaping table will be indexed using IQ vector magnitude (0...1) and the ET DAC value. It is possible to index the shaping table using either linear or log magnitude, which will influence the number of entries required.

Since the shaping table could be updated based on changes in channel frequency, power control level, temperature, or changes in PA load impedance, the hardware implementation of the LUT should be double-buffered to support "atomic" updating of all values at an LTE TTI boundary.

10.2.7   Post-LUT Digital Gain/Offset Correction

Following the shaping table, a further digital gain and offset control is used to compensate for any part-to-part variations in analog gain or offset of the ET DAC or ETIC.

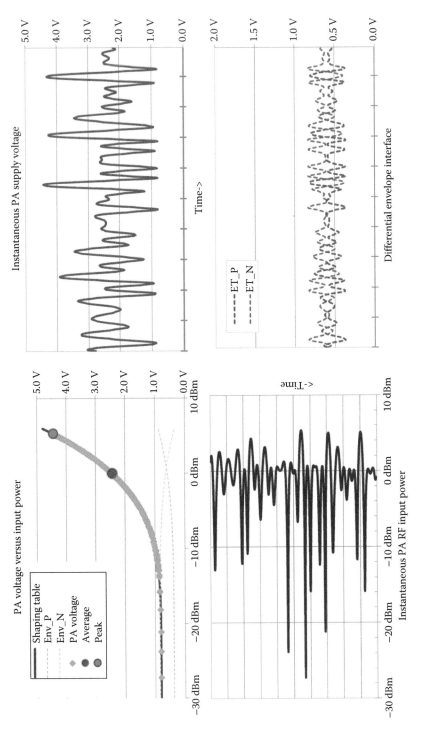

FIGURE 10.4 ET signal waveforms, 0 dBm at the PA input.

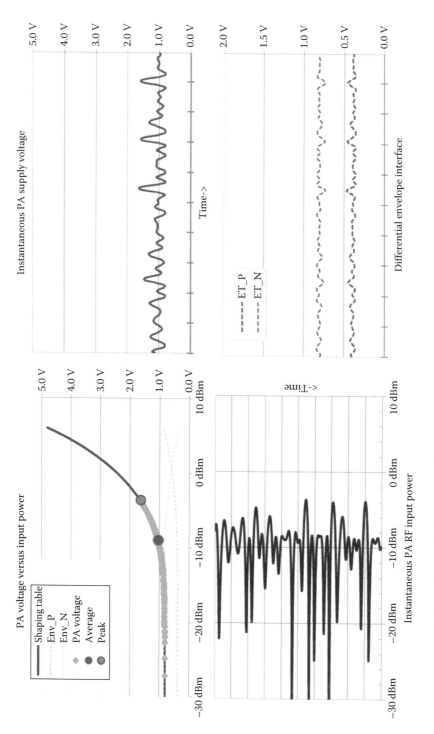

FIGURE 10.5 ET signal waveforms, −9 dBm at the PA input.

The nominal voltage gain and offset (output voltage with zero differential input) of the ETIC is specified by each manufacturer and is not standardized.

The ET path gain and offset vary from part to part due to analog variation in the ET DAC, reconstruction filter, and ETIC; since the overall contribution will be a function of both the modem chipset and ETIC, this requires trim or calibration at production time. Depending on the behavior of the ETIC and modem chipset over temperature, some temperature compensation of these parameters may also be required in "mission mode."

10.2.8 PRE-DAC UPSAMPLING

Depending on the sample rate chosen, a further interpolation stage may be implemented just before the DAC in order to reduce the aliasing components.

In architectures which scale down the ET path clock rate as the LTE channel bandwidth reduces, this can be particularly important, as the DAC images may then fall within the bandwidth of the ETIC—if an alias component falls within the FDD RX band, this could lead to significant desensitization.

10.2.9 ET DAC REQUIREMENTS

The ET DAC is usually one of the "pinch points" in ET system design, due to limitations on available sample rate, resolution, or both. Most LTE implementations use current-steering DACs with on-chip resistors to convert from current to voltage.

Typical DAC specifications for 20 MHz FDD-LTE applications are 10-bit ENOB and 240 MSPS, which provides a good compromise between power consumption and performance.

Reducing the DAC resolution will increase the quantization noise floor on the ET path, which will contribute to RF noise levels at the PA output. Reducing the sample rate may cause problems with ET path image frequencies falling into the RX band—due to the large number of FDD band frequency offsets and possible channel bandwidths, this is a complex system analysis to perform.

10.2.10 POST-DAC VOLTAGE BUFFERING AND FILTERING

Some chipsets have integrated on-chip voltage buffers after the ET DAC, but these have challenging noise requirements, which can drive up power consumption. The latest V1.1 release of the MIPI eTrak specification provides for a lower common-mode voltage option, which is better matched to the lower supply voltages of deep submicron modem chipsets, and should allow for the direct drive of an ETIC from an on-chip DAC.

Similarly, some modem chipsets and ETICs provide switchable analog bandwidth filters—these are primarily of use in suppressing DAC images if the modem clock rate is scaled with LTE channel bandwidth. Although these filters could be implemented on the modem chipset, it is generally preferable to integrate them in the ETIC, where they can also filter out noise generated in the early stages of the ETIC and any noise or interference generated by the eTrak interface.

The final interface to the ETIC is a differential voltage-mode signal; the MIPI eTrak specification provides for three different classes with different maximum swing ranges and common-mode voltage ranges.

10.3   MODELING THE ET PA

When designing and optimizing an ET system, having a good simulation model of the PA is essential. The gain characteristic (AM/AM response) and phase characteristic (AM/PM response) of a PA are the primary distortion mechanisms which cause degradations in EVM, ACLR, or out-of-band noise.

A traditional fixed-supply PA can be modeled as a two-port device (RF in/RF out), making it relatively easy to characterize and model the amplifier's linearity. The PA's gain and phase responses depend only on the instantaneous RF input amplitude. It is, therefore, straightforward to produce a figure-of-merit for the intrinsic linearity of the PA, for example ACLR or EVM with a specified waveform at a particular mean output power.

An ET PA must, however, be modeled as a three-port device, with two input ports (RF and Supply), and one output port (RF out). The AM/AM and AM/PM responses of the PA depend on both the RF and supply inputs.

It is important to understand that, when considered as a standalone device, the ET PA has no intrinsic "linearity." It is only when the ET PA is combined with a specific shaping table and reverts to a two-port device that linearity becomes meaningful—the linearity is as much a function of the shaping table as it is of the PA's intrinsic device characteristics.

The ET shaping table defines the instantaneous supply voltage for any given input power. The combination of the PA characteristics and a particular shaping table together define the AM/AM and AM/PM responses, and hence the overall system linearity. Whereas the shaping table directly influences the AM/AM response, the AM/PM response is primarily determined by the PA's intrinsic characteristics.

The combination of three-port ET PA plus shaping table can, therefore, be modeled as a two-port device, with the AM linearity defined by the shaping table. It is this aspect of ET that (to some extent) makes the ET PA a "software-defined PA," where the shaping table can be used to trade-off linearity and efficiency.

10.3.1   CHARACTERIZING THE ET PA

As we have seen, the "standalone" performance of ET PAs cannot be measured unless the shaping table is first defined. This requires measurement of the PA's fundamental characteristics (output power, efficiency, gain, and phase) over the full range of instantaneous supply voltages and input powers. There are several methods for ET PA characterization, summarized in Table 10.1.

In principle, the characterization could be carried out using a CW network analyzer and a variable DC supply, but results are typically poor due to thermal effects, ranging errors, and drifts in phase measurements. It is also far too slow to allow load-pull techniques to be used.

TABLE 10.1

ET PA Measurement Methods

Test Methodology	PA Current Measurement	Supply Impedance	Supply Bandwidth Requirements	ET Efficiency Prediction	ET Linearity Prediction	Parameters Measured
Swept CW testing	Bench PSU	Low (decoupling capacitor)	Low (bench PSU)	Poor, due to PA die heating	Poor, due to PA die heating	Gain (AM:AM), efficiency
Pulsed RF/DC testing	Instrumentation grade current probe, ~5 μs resolution	Low (decoupling capacitor)	Low (bench PSU)	Good, if short pulses (~10 μs, 10% duty cycle)	Fair	Gain (AM:AM), efficiency
Dynamic supply modulation	Challenging—high BW with high common mode voltage current sense	Requires low impedance dynamic supply (no decoupling)	High (~60 MHz BW)	V. Good	V. Good	Gain (AM:AM), phase (AM:PM), efficiency

An alternative approach is to use RF pulse characterization using ATE-controlled standard test equipment. This avoids the need for a high-bandwidth, low-impedance supply and is sufficiently fast for load-pull to be viable, but has the drawback that it is difficult to make accurate phase measurements.

The last approach is to use real ET waveforms and to dynamically vary the shaping table to allow all combinations of input power and supply voltage to be measured. This requires a high-bandwidth supply modulator, but is very fast, allows accurate phase information to be gathered and can in principle also be used to characterize memory effects.

This basic ET PA characterization can be used to create a quasi-static (i.e., "memoryless") data model of the PA having output power, phase and efficiency as outputs, and input power and supply voltage as inputs. Once the shaping table is defined, this model can then be used to predict PA system performance parameters, such as ACLR, EVM and efficiency, for standard test waveforms.

In addition to being used for PA device level characterization, the same hardware can be used for direct verification of PA system performance using the defined shaping table.

For higher bandwidth waveforms, memory effects can be a significant source of nonlinearity. The PA output parameters (AM, PM, and efficiency) now depend on time (i.e., signal history) in addition to instantaneous input power and supply voltage. Memory effects show up in the PA characterization as a "broadening" of the AM/AM and AM/PM characteristics and can result from electrical time constants in input or output bias circuits, thermal time constants associated with local die heating, or technology-specific "charge storage" effects. For low power handset PAs, memory effects associated with the gate/base bias and drain/collector feed tend to dominate over die level memory effects.

10.3.2   THE 3D SURFACE MODEL OF THE ET PA

By post-processing data captured over a range of supply voltages and RF input powers, it is possible to analyze the efficiency, gain, and phase of the PA as a set of 3D surfaces or parameterized 2D plots, enabling the PA designer to visualize the amplifier characteristics and make suitable design trade-offs during performance optimization (Figure 10.6).

The same data set can also be used to automatically calculate and export a variety of shaping table mappings to target maximal efficiency, maximal linearity, or other target performance parameters. During shaping table design, it can be helpful to view the gain, phase, and efficiency surfaces in different orientations to visualize how the selected shaping table will impact AM:AM, AM:PM, and efficiency responses, as shown in Figure 10.7.

The left-hand plot in Figure 10.7 shows the parameterized 2D "traditional" view of the surface, showing gain (AM/AM) on the Y axis, and the variable supply voltage shown as a series of curves. The operating locus with the selected shaping table is highlighted in black—in this case a 25.5 dB isogain shaping.

The right-hand plot shows the "shaping table" view of the surface with instantaneous supply voltage on the Y axis and instantaneous PA input power on the X axis.

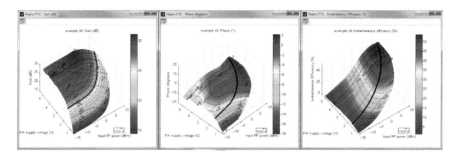

FIGURE 10.6   3D surfaces of the ET PA gain, phase, and efficiency.

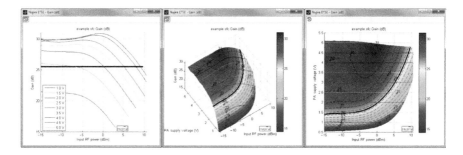

FIGURE 10.7   Alternative views of the ET PA gain surface; the ET shaping table is high-lighted in black.

The color map (Z axis) represents gain, with the contours of equal (iso) gain high-lighted, and the selected shaping table again highlighted in black.

Figure 10.8 shows gain, phase, and efficiency surfaces side-by-side, with the tra-ditional view at the top, and the shaping table view of the surface below.

The 3D "quasi-static" surface model of the PA can also be used in a simulation environment with the selected shaping table—by applying a sequence of IQ samples for a defined waveform at a selected mean power level at the PA input, the model can be used to accurately predict ACLR, EVM and mean efficiency at a system level across a wide range of waveforms and power levels, provided memory effects are sufficiently well controlled.

10.3.3   Noise and Distortion Mechanisms in ET

When operating in ET mode, the PA is in compression and, therefore, acts as a mixer—noise and distortion generated in the ET supply path mixes with the RF signal and increases noise and distortion at the PA output.

The amount of supply noise transferred to the RF spectrum is a function of the instantaneous compression level of the PA—when in saturation, the ET PA has no power supply rejection, whereas in linear operation, it has significant inherent rejec-tion of supply noise. Like many other aspects of ET PA performance, the supply

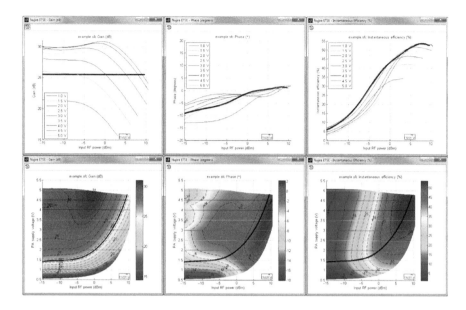

FIGURE 10.8 Gain, phase, and efficiency surfaces.

noise sensitivity of the PA can be controlled by the shaping table—operating the PA further into compression increases efficiency, but also increases the supply noise transfer, and vice versa.

It is, of course, important to minimize the analog noise and distortion present on the ET supply pin of the PA, but at a wider system level there are multiple interacting sources of noise and distortion to be considered, which may be generated by digital, analog, or software-controlled mechanisms.

Some contributors—for example, insufficient ET DAC precision or switcher noise—increase the broadband noise floor. Other mechanisms, such as ET/RF path gain mismatch or timing alignment error, introduce degradations close to carrier and appear more like traditional distortion due to PA compression. Insufficient sample rate in the ET reference signal calculation or inadequate ETIC stability margin may give rise to "humps" in the RF spectrum either side of the wanted channel.

Depending on the system requirements, the PA performance limits may be dictated by in-band performance (e.g., EVM for high-order QAM modulation) or out-of-band requirements, such as ACLR, FCC regulatory limits, or self-interference (TX path noise in the RX band).

For FDD-LTE handset applications, RX band noise is usually the limiting factor for ET path performance—unwanted noise or distortion at the PA output is attenuated by the duplex filter, adds to the wanted RX signal from the antenna, and degrades the receive sensitivity of the handset.

TX/RX frequency band separation varies significantly by LTE frequency band, so the relevant noise products can be as close as 30 MHz to the TX carrier, or up to 400 MHz away depending on the band of operation. The baseband spectrum of the ET path must, therefore, be considered carefully across a wide frequency range, as an unwanted noise spike—for example, switch-mode power supply noise, or ET DAC image frequency—may fall directly into an FDD receive band.

With so many potential sources of noise and distortion, and a wide range of use cases across LTE frequency bands and channel bandwidths, a high-level system simulation environment is very helpful. Such models can be constructed in environments, such as MATLAB or ADS.

To isolate and examine each individual noise and distortion contributor, a very high dynamic range three-port PA simulation model is needed—ideally with around 150 dB of dynamic range. Although it is possible to generate such a model by interpolating lab measurements of the PA, an equation-based model is generally preferable as it can provide smooth, continuous surfaces, which allow simulation of contributors at 130 or 140 dB below the transmission power level.

10.4 ET SHAPING TABLE

As already discussed in an ET system, the behavior of the PA is defined by the contents of a nonlinear envelope "shaping table" in the digital signal processing path which generates the envelope reference signal. The shaping table, which maps the instantaneous RF output power to supply voltage, is specific to the type of PA being used and the system performance requirements.

Although it is possible to create a unique shaping table for each individual PA at production time, it is simpler to fix the shaping table at design time, and calibrate out just the analog gain and offset errors in the envelope and RF paths on the production line.

Depending on the PA being used, a different shaping table may be needed for each frequency band, for example in the case of a multimode, multiband PA. Better performance may also be obtained by generating multiple shaping tables to cover bands with particularly wide fractional bandwidths.

10.4.1 INFLUENCE OF THE SHAPING TABLE

In most applications, the PA has to simultaneously meet several key metrics— efficiency, output power, gain, in-band linearity (EVM), and out-of-band linearity (ACLR and RX band noise), potentially over a wide power control range and a wide frequency range. These metrics are all interlinked and improving the performance in one area means trading off the performance in another. For example, PA efficiency and gain both vary with supply voltage—a higher supply voltage gives you more gain through the PA, but lower efficiency, and vice versa.

"Optimizing" an ET PA design is, therefore, a process of finding the right balance between these metrics to give the best overall system performance.

With fixed supply PAs, product designers have had very limited control over these metrics. ET transforms the PA from a two-port into a three-port device, with the supply pin acting as an additional high-bandwidth control input, giving designers more freedom to define the trade-offs required to optimize PA performance.

At each instantaneous power level, a higher supply voltage will give more gain (at the expense of efficiency) and a lower supply voltage will give more efficiency (at the expense of gain). This technique can be used in the shaping table to control the desired AM:AM characteristic of the PA, and hence the linearity.

When the ET PA is operating in compression/saturation at high instantaneous power, increasing the RF input amplitude does not provide any more output power— that is the definition of saturation. The only way to get more output power from a PA in saturation is to increase its supply voltage. In other words, the output amplitude is directly controlled by the supply voltage.

The profile of the ET shaping table controls how the supply voltage varies with PA input signal amplitude. The AM:AM distortion of the PA (i.e., how its output signal amplitude changes as a function of the input signal amplitude) is, therefore, directly controlled by the shaping table. A "kink" in the shaping table will give a corresponding "kink" in the AM:AM response and degrade linearity.

Phase is not controllable by the supply voltage in the same way as the gain of the PA, but the phase response (AM:PM distortion) of the PA in ET mode can certainly be characterized, and to some extent controlled, via the shaping table—but this also depends on the design of the PA bias and matching networks.

The ability to directly control and trade-off gain and efficiency, and indirectly control phase, allows the shaping table to control a wide range of PA performance characteristics, as listed in Table 10.2.

10.4.2 Isogain Linearization

One obvious ET shaping table is the maximum efficiency shaping, where the supply voltage is selected to maximize the instantaneous efficiency of the PA at all signal levels. However, with this approach, the PA gain is usually nonlinear, resulting in significant AM:AM distortion as shown in the left side of Figure 10.9. This response would require correction with digital predistortion (DPD) to achieve the system linearity requirements. However, the main disadvantage of this approach is that high-bandwidth predistortion would be needed to effectively correct the sharp "kinks" in the AM:AM characteristic. In addition to the baseband processing

---

TABLE 10.2
System Parameters Controlled by the Shaping Table

Primary PA Performance Parameters—Directly Controlled by the Shaping Table	Secondary System Performance Parameters—Influenced by the Shaping Table
PA efficiency	ETIC efficiency
AM:AM linearity	AM:PM linearity
Average PA gain	Envelope signal bandwidth
	Waveform PAPR at PA output
	PA noise transfer/supply sensitivity
	PA peak and minimum supply voltages
	Isogain power control range
	PA supply port impedance/linearity

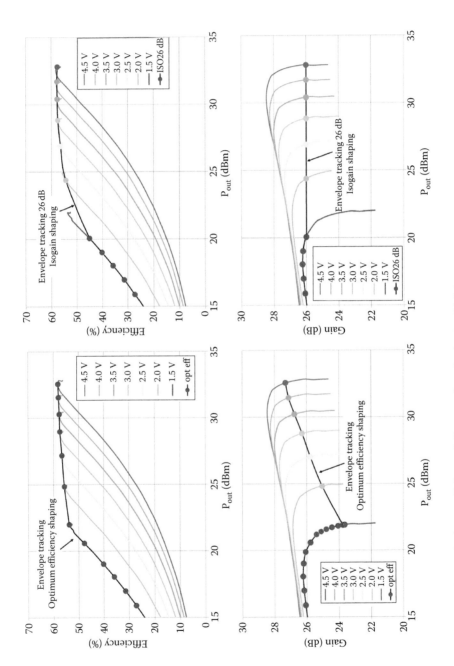

FIGURE 10.9 PA efficiency: Optimum efficiency (left) versus isogain shaping (right).

implications of the required high bandwidth, it also restricts the ability to filter out RF path noise.

An alternative mapping of particular interest is "Isogain" shaping in which the instantaneous supply voltage is chosen to achieve a particular constant PA gain, as illustrated on the right-hand column of Figure 10.9.

With isogain shaping, the ET PA system achieves low AM:AM distortion, while still operating in compression over much of the modulation cycle. For typical GaAs LTE handset PAs, the efficiency penalty for using isogain shaping rather than max efficiency shaping is often as low as 1% or 2% points, but the ACLR can be 7–10 dB better.

Depending on the PA design, it is possible to use isogain shaping to linearize the PA down to ACLR values of −45 dBc or even better without requiring any form of DPD.

It is important to note that there are an infinite number of potential isogain shaping tables to choose from—selecting the best isogain value is a trade-off between efficiency, linearity, and the average power control range over which the isogain remains valid.

To maintain the same isogain over a range of power control levels requires an isogain value which maximizes the flat (controlled) area of the gain characteristic. For example, in Figure 10.9, it can be seen that with the highlighted 26 dB isogain trajectory, a slight "kick-up" in the gain occurs at low instantaneous power levels as the PA drops out of compression. This introduces some AM:AM distortion in the troughs of the signal, which may not be significant at high average power levels, but will degrade ACLR as the average power level is reduced. By contrast, choosing a slightly higher isogain, for example, 26.2 dB, would allow isogain operation to be maintained at a lower average output power, at the cost of a slight decrease in efficiency.

10.4.3   OPTIMIZING THE PA PHASE RESPONSE FOR ET

Since the shaping table can be used to directly control the AM:AM behavior of the PA in ET mode, the residual distortion is generally due to AM:PM, which can be relatively easily corrected using open-loop phase precorrection in the IQ path.

Alternatively, the PA's phase response can be designed to minimize AM:PM distortion in ET mode, as illustrated by the two phase surfaces shown in Figure 10.10.

The PA on the left has been designed for average power tracking (APT) operation and has a relatively flat AM:PM response when modulated with a fixed supply. However, in ET mode, the supply voltage trajectory crosses a significant number of phase contours, generating some relatively significant AM:PM distortion.

The surface shown on the right is taken from an ET-optimized PA where the phase response has been optimized along the ET supply voltage contour, giving an extremely flat AM:PM response when operated in ET mode, but significant AM:PM distortion when operated in APT mode. ET-optimized PAs can achieve ACLR performance in excess of −50 dBc without DPD. This significantly exceeds the LTE uplink requirement, so usually the shaping table is chosen to achieve the best efficiency consistent with meeting linearity goals.

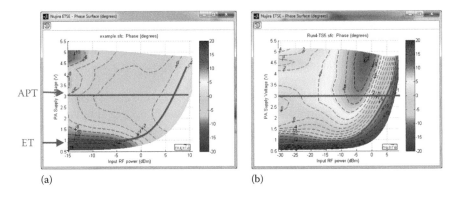

APT

ET

(a)   (b)

FIGURE 10.10   PA phase surfaces for APT-optimized PA (a) and ET-optimized PA (b).

10.4.4   APPROACHES TO SHAPING TABLE DESIGN

There are several methods which can be used to develop ET shaping tables. The surface-based PA characterization method described can be used to automatically calculate and export shaping tables by curve fitting the shaping table to measured gain, phase, or efficiency surfaces. This approach requires an accurate ET PA characterization bench and must ensure that the thermal profile of the PA during surface capture accurately reflects the dissipation when operating in ET mode.

Alternatively, an equation-based approach to shaping tables can be used, based on the peak and minimum ET voltage, the peak PA input power, and some other parameters—for example, to control sharpness of the transition from linear operation to compression.

Equation-based shaping tables have the advantage of simplicity and a small number of parameters and can be relatively easy to develop in the lab, particularly if closed-loop DPD is available, or if ACLR targets are not too stringent.

When designing a shaping table, there are several factors to consider.

The peak ET voltage is determined by the peak power required at the PA output and should be regarded as "fixed." Peak output power is, in turn, defined by the minimum PAPR needed at the PA output to achieve the system linearity targets and must also take into account the peak ET supply voltage supported by both the ETIC and PA.

The minimum ET voltage used in the troughs is a design trade-off between PA efficiency, ETIC efficiency, ET bandwidth, and linearity. For GaAs HBT PAs, it is advisable to keep the minimum voltage above around 1.0 V to avoid gain collapse in the PA. Increasing Vmin reduces PA efficiency, but increases ETIC efficiency by reducing the AC power delivery requirement. It also reduces the ET bandwidth required. The opposing PA and ETIC efficiency trends result in a broad efficiency "sweet spot" for swing ranges in the range 2.5:1–3.5:1 (e.g., Vmax = 3 × Vmin). Note that the "optimum efficiency" swing range may not always deliver the best system linearity unless AM precorrection is used.

It is important to focus on optimizing efficiency at the maximum mean output power level, by targeting a shaping table voltage which gives 2–3 dB compression at that point—this gain being a good starting point for deriving an isogain shaping table. Although the mean efficiency depends on the statistical distribution of the signal waveform around the mean power level, the spot efficiency at that power level is usually a reasonable proxy. Note that the PA will normally operate deeper into compression at the waveform peaks—perhaps up to 4 dB—but by even more if the shaping table is also used to provide deliberate "soft clipping" of the waveform.

The influence of the shaping table on the ET-path bandwidth can also be significant. Any "sharp corners" in the shaping table, or non-monotonic sections, can cause a significant increase in the ET-path bandwidth, and lead to mistracking if the capabilities of the ETIC are exceeded. The minimum shaping table voltage also serves to "de-trough" the envelope waveform, significantly reducing the envelope path bandwidth requirement, since the unshaped envelope has very sharp troughs where the RF envelope falls to zero, as illustrated in Figure 10.11.

10.5   ET SUPPLY MODULATOR

The ET supply modulator is both a power supply and an amplifier; at a functional level, it amplifies the differential MIPI eTrak envelope reference signal provided by the modem chipset to supply the PA, and for simple, system modeling can be considered as an ideal amplifier with a low-pass filter plus an additive noise source.

The engineering challenge in developing an LTE ET chip is in creating a power supply converter that simultaneously achieves high efficiency (>80%), extremely high bandwidth (30–60 MHz for 20 MHz BW LTE), very low noise, and high peak output currents (>1 A).

For IC designers, delivering an effective ET power supply modulator requires an architecture that delivers the right balance between the key metrics—system efficiency, noise and distortion, bandwidth, and output power—while paying attention to some of the less obvious requirements, such as output impedance—which influences switcher noise suppression and supply "memory effects," slew rate—which can result in ETIC mis-tracking—and mode transition times.

LTE ET chips must also support fallback to traditional average power tracking operation at reduced output powers, where they must deliver competitive efficiency and ideally consume no more current than an APT-only DC:DC converter when heavily backed off.

10.5.1   SMPS-Only ETIC Architectures

The simplest ETIC architecture variant is that of a standalone fast switch-mode power supply (SMPS, often referred to as a DC:DC converter). Switcher-only architectures are attractive from a size, cost, and efficiency perspective.

The main drawback of switcher-only architectures is that it is very difficult to deliver the high bandwidth needed without compromising efficiency. As a rule of thumb, a ratio of around 30:1 is required between the switching frequency and the RF channel bandwidth. To support an LTE channel bandwidth of 20 MHz would, therefore, require a switching speed of at least 150 MHz, or use of multiphase switching converters to achieve the required bandwidth.

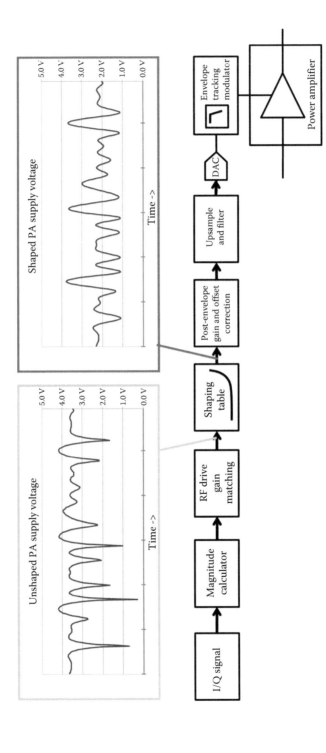

FIGURE 10.11 Envelope bandwidth before and after shaping table.

Suppressing switching noise is also an important consideration—each switching event generates a noise spike, which may result in spurs being transferred through the PA onto the RF spectrum. This may require relatively heavy filtering on the ET supply, limiting bandwidth and efficiency.

At high switching speeds, parasitic losses in the switching devices and inductors can also become significant, resulting in an efficiency falloff as switching frequency increases. At backed off power levels and APT modes, it is usually preferable to use a slow-switching architecture, avoiding the standing current penalty of the fast switcher, but this is hard to achieve with a single inductor value without resorting to discontinuous switching, which decreases supply bandwidth and increases LF noise.

10.5.2 Hybrid ETIC Architectures

It is possible to overcome the challenges of SMPS-only ETICs by adding a high-bandwidth linear (AC) amplifier to the ETIC to create a "hybrid" architecture. This is conceptually similar to multiple loudspeakers in an audio system, where a "woofer" handles the high-power/low-frequency content, and a smaller "tweeter" provides the high-frequency content (Figure 10.12).

In hybrid ET architectures, the switch-mode power supply handles the DC and low-frequency components of the ET supply waveform, which contain most of the energy consumed by the PA (typically 85%–95%, depending on the waveform statistics). The switch-mode power supply can achieve very high energy conversion efficiency—90%–95% is a good benchmark—and, in the hybrid architecture, does not require an excessively high switching rate.

The AC amplifier then provides the high-frequency components of the ET supply waveform, delivering the high signal bandwidths required and—depending on the details of the ETIC architecture—cleaning up noise generated by the switch-mode power supply. The AC amplifier is usually biased in a class AB configuration, which

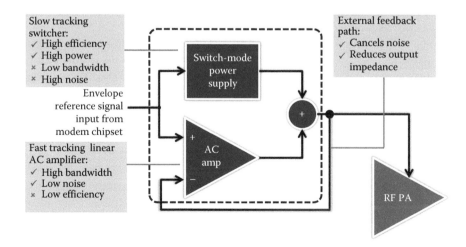

FIGURE 10.12   Hybrid ETIC architecture.

can only deliver theoretical efficiencies of around 40%–50% for the AC amplifier itself. However, since only a small portion of the energy supplied to the PA is delivered by the AC amplifier—perhaps only 5%–10%—the effect on the overall energy efficiency of the ETIC is not dramatic.

10.5.3   ETIC FIGURES OF MERIT

It is not straightforward to compare ETICs, as there are multiple figures-of-merit to consider, including

- Maximum ET bandwidth
- ET voltage swing range
- Efficiency with high-bandwidth signals
- Efficiency with low-bandwidth signals
- Efficiency over back off
- Output impedance
- Current and voltage slew rates
- Buck, AC-boost, or DC-boost capability
- Supply voltage range
- Ease of programming

10.5.4   ETIC BANDWIDTH

One of the main implementation challenges for an envelope tracking IC is that of bandwidth—the power supply requires a small signal bandwidth of 1.5–3.0× the RF channel bandwidth, that is, 30–60 MHz for a 20 MHz OFDM channel.

The spectral plot of the envelope signal power shows that there is a large DC component, followed by a significant amount of power within the occupied channel bandwidth, and then a relatively long "tail" out to about 3× the channel bandwidth. How much of this tail must be accurately tracked by the ETIC depends on the system RF linearity requirements.

Some ETICs incorporate switchable bandwidth filters to support different channel bandwidths—these can be helpful at lower channel bandwidths to attenuate ET DAC images, particularly if the modem chipset reduces the ET DAC clock frequency at lower LTE channel bandwidths.

10.5.5   ETIC EFFICIENCY

Perhaps harder than achieving the absolute bandwidth through the ETIC is to maintain high energy conversion efficiency for the ETIC itself. While 85% ETIC efficiency is a good figure of merit for low-bandwidth signals such as 3G, maintaining this efficiency over a wide range of signal types and bandwidths is a significant architectural and IC design challenge.

Some ETICs are optimized for efficiency at a low bandwidth, for example, 1–4 RB, with a significant falloff in efficiency at higher RB counts. Others may provide a lower efficiency for low bandwidths but maintain efficiency better as the bandwidth increases.

Maintaining high efficiency over a wide power control range is also important for applications, such as LTE and 3G, that employ dynamic power control. ET usually delivers a power saving benefit over the top 8–10 dB of the power control range and the efficiency of the ETIC as power is backed off usually determines the "break-even" point compared with APT.

ETICs for cellular applications typically also include a lower bandwidth APT mode for operation at reduced power levels, and this must also operate at high efficiency over a very wide power output range—from 10 mW to more than 1 W—making circuit design for low quiescent currents an important architectural and design consideration.

10.5.6 Low Noise

For FDD-LTE systems, analog "noise" created by the ETIC is an important consideration, as it is one of the contributors to overall RX sensitivity degradation. However, not all "noise" is equal—there are multiple sources of noise and distortion within the ETIC, all of which contribute to the overall system noise budget.

Some contributors can be considered as true noise, such as thermal noise in the ETIC amplifier and switching noise from the SMPS blocks. It is not possible (or practical in the case of switching noise) to predict and precorrect the RF signal to mitigate these sources of noise.

Other effects, such as crossover distortion, bandwidth limitations, high output impedance, or slew-rate limitations, lead to mistracking or a "noise-like" distortion of the ET signal. Some of these effects give rise to deterministic memory effects, which could, in principle, be corrected using memory effect predistortion, though this is not typically used in current handset implementations.

10.5.7 Slew Rate

In addition to a high small-signal bandwidth, ET modulators also require a high slew rate in order to accurately track the peaks and troughs of the waveform.

To accurately track a 20-MHz LTE uplink envelope with a 3-V swing at typical handset output power levels needs over a 150 V/μs voltage slew rate, requiring the output current to slew at about 50 A/μs.

Shortfalls in the slew rate will cause mistracking, leading to increased noise and distortion on the RF output—this may be particularly harmful in FDD systems where transmit noise in the receive band can degrade the receiver sensitivity.

Although the required voltage slew rate can (to some extent) be controlled at the system level by the choice of the shaping table, the dynamic behavior of the PA current consumption is determined primarily by the instantaneous RF output power characteristics, and cannot easily be reduced. This tends to lead to high-bandwidth current "spikes" and "clicks" caused by rapid changes in the RF output amplitude.

10.5.8 OUTPUT IMPEDANCE

If the ETIC has high output impedance, this results in voltage errors due to the fast current changes in the load, leading to distortion and noise on the RF output. Typically, the output impedance of a feedback amplifier is primarily inductive and hence rises rapidly with frequency. Consequently, the ability of the correction amplifier to "clean up" residual switcher noise falls with frequency. Furthermore, the inductive output impedance together with the inductance associated with the ETIC—PA physical interconnect results in memory effect distortion, which may become significant with very high-bandwidth signals.

The ETIC effectively replaces the supply decoupling capacitor on the PA at video frequencies, so low ETIC output impedance is also vital to maintain PA stability. ETICs will typically be connected to multiple PAs on the board, which add parasitic capacitance and stray inductance even when the unused PAs are turned off. Resonances associated with these parasitic elements are difficult to predict and control and can result in higher than expected out-of-band noise.

RF PAs present a time varying load impedance to the ETIC, making them far more complex to drive than a simple resistive load. The nature of this "dynamic" impedance depends on the PA technology and the choice of shaping function, and typically rises in the waveform troughs as the PA comes out of saturation. The rapid change of impedance can give rise to transient mistracking or "trough clicks" in waveform troughs. It also requires more power to be delivered by the linear amplifier of a hybrid ETIC than when driving an equivalent "linear" resistance—resulting in the ETIC efficiency reducing by about 2 percentage points.

10.5.9 "AC BOOST" VERSUS "DC BOOST"

Although portable products use batteries with 3.8 V nominal voltage, designers are increasingly looking for components including PAs, which operate down to 2.8 V (or even 2.5 V) supplies. Although it is possible to design low load line PAs capable of max power operation at these voltages, several key PA parameters must be compromised to do so. In particular, the RF bandwidth and PA efficiency are both degraded and the PA's peak supply current is increased.

A common requirement is, therefore, for the ET modulator to be able to deliver peak voltages of 4.5–5 V by boosting above the battery voltage, although the mean supply voltage to the PA in ET mode is almost always below the battery voltage (typically 2–2.5 V).

This can often be achieved through "AC boost"—the ability of the linear amplifier within the ETIC to deliver transitory peak output voltages above the supply voltage. This allows full power ET operation even at low battery.

However, this precludes use of APT at full power, which may be required to support legacy standards, such as 3G or TD-SCDMA, where the modem chipset vendor is unwilling to add ET support. AC boost may also result in degraded low-bandwidth LTE performance. Where this is a problem, "DC boost"—the ability for the switch-mode power supply within the ETIC to sustain an output voltage above that of the supply—may be required.

Using DC boost, ETICs may degrade efficiency compared with an AC boost architecture, due to the increased complexity of the switch-mode power supply block, though the extent of degradation is strongly architecture dependent.

10.5.10   ET/APT Mode Transitions

In LTE systems, transmit power control at the handset depends both on distance from the base station (path loss) and the instantaneous bandwidth being transmitted during each timeslot, which can result in a 100× variation in transmit power each millisecond.

The ETIC, therefore, needs to be able to alter transmit power level, and switch between ET and APT modes in a few microseconds without introducing "glitches" into the supply of the RF PA—which may cause in-band (EVM) or out-of-band (ACLR) distortion during the mode transitions.

In APT mode, the ETIC typically maintains low output impedance by switching in an appropriate decoupling (bypass) capacitor across the output supply to reduce ripple and maintain PA stability. This switching must also be carried out while minimizing disturbance of the output signal.

10.6   HOW ET AFFECTS THE PA DESIGN

The requirement to measure, characterize, and model the ET PA as a three-port device has already been discussed, but at a hardware level, there are several practical differences between an APT-optimized PA and an ET-optimized device.

To make a PA "ET capable" requires at least the following modifications:

- Removal of any large decoupling components on the final stage supply ("VCC2"), leaving only RF decoupling
- Ensuring that any internal bias circuits are not powered from the VCC2 pin—ET PAs typically provide a separate Vbat supply input for this purpose
- Supporting modulation of the VCC2 pin over at least a 3:1 range, for example, around 1.2–3.6 V

To optimize a PA for ET operation, there are several further design techniques that can be employed:

- Increasing the PA load line to match the "AC boost" capability of the ETIC
- Supporting ET co-modulation of the PA driver stage ("VCC1")
- Optimizing the phase response for ET rather than APT
- Ensuring the bandwidth of the supply bias networks is sufficient for ET modulation
- Optimizing efficiency around mean rather than max power
- Ignoring any requirement for fixed supply AM:AM linearity at high power
- Ensuring the PA gain falls off as the supply voltage reduces, for good isogain
- Reducing the static PA bias current/Icq (fixed supply gain peaking is not an issue for ET operation)

10.6.1 PA Supply Decoupling

One obvious consequence of ET operation is that, unlike an APT PA, the PA's VCC2 supply must not include any significant decoupling at video frequencies. RF decoupling is still required, but is typically 100–150 pF per cellular PA for multi-PA applications. Note that the ETIC still sees the "off" capacitances of the inactive PAs, and the presence of these load capacitances can cause mistracking, out-of-band resonances, and increased power consumption, particularly with high-bandwidth waveforms.

10.6.2 ET Swing Range, PA Load Line, and Supply Impedance

The load line of an ET PA and the ETIC swing range should be chosen to match one another. The load line is defined by the peak PA output power required and the peak ET voltage the ETIC/PA combination can support. For an LTE PA capable of delivering 28 dBm of average power with a 5 dB PAPR, paired with an ETIC/PA capable of 4.5 V peak voltage, the PA load line should be optimized for 33 dBm pk power.

The term "load line" is used rather loosely with ET PAs, and refers to the PA's "supply impedance"—the relationship between instantaneous supply current and voltage on the ET-modulated supply pin of the PA (as distinct from the RF load line, which is the relationship between the PA's RF voltage and RF current). For example, delivering 33-dBm peak power from a 4.5-V peak ET supply voltage, assuming 60% saturated PA efficiency, implies a supply impedance of 6 ohms. Delivering the same output power from a 3.6-V peak supply implies a 4-ohm PA supply impedance.

PAs designed for APT operation are usually designed to deliver full power from a buck-only DC:DC converter, resulting in a relatively low load line/supply impedance (3.5 or 4 ohms). When used with a boost-capable ET supply, these devices can typically deliver 2–3 dB more linear power than in APT mode, due to the higher peak supply voltage available.

Higher load-line PAs are generally more ET-friendly and can make use of higher peak voltages to deliver the increased peak power. The challenge of high load-line PAs is that if APT operation at full power is also required—for example, to support legacy 3G standards—they require a boost-mode APT converter to deliver full power.

10.6.3 Driver Stage Modulation

The power consumption of the PA driver stage can become a significant factor in ET PA designs, where the final stage efficiency may be reaching 60%. As the gain of the final stage is reduced when operating in ET mode, an additional 1–2 dB of linear power may be required from the driver stage leading to an increase in energy consumption compared to an APT design.

For this reason, a simple battery-connected supply for the driver (VCC1) stage is not ideal, particularly as power is backed off in APT mode. Options include provision of a dedicated APT supply rail for the driver stage or co-modulation of both the driver and final stages.

If co-modulation is used, additional gain must be provided in the PA to compensate for the compression in the driver stage. Care must also be taken to when designing the PA to optimize the AM/PM response across both stages and to avoid memory effects introduced by the propagation delay through the driver stage.

10.6.4 OPTIMIZING THE PA FOR ET LINEARITY

If the PA is not going to be used in APT mode at full power, it is possible to increase the efficiency by sacrificing some intrinsic linearity. The ideal ET PA gain characteristic is shown in Figure 10.13.

At high instantaneous powers, the fixed supply gain flatness of the PA as it approaches compression is unimportant, as the AM:AM response in this region is defined by the ET shaping table. The bias current can therefore be reduced, and the gain allowed to rise as the PA approaches compression.

At low supply voltages, shown as V1 in Figure 10.13, the PA gain should be allowed to reduce, as this provides a greater dynamic range over which the isogain profile is flat. At lower instantaneous powers, the PA should be designed for intrinsic linearity, as the ET shaping table will not have any effect in this region.

10.6.5 OPTIMIZING THE PA FOR ET EFFICIENCY

The main efficiency goal for an ET PA is to ensure that the saturated efficiency remains high as the supply voltage is backed off, as illustrated by the curves in Figure 10.2.

The most important operating point is the region around the mean output power level, that is, 5–7 dB below peak power for LTE uplink signals, where the ET supply voltage will usually be around 1.5–2 V. This is more important than the saturated efficiency at peak power.

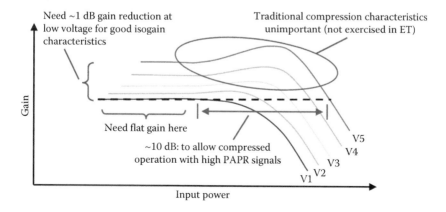

FIGURE 10.13   Ideal ET PA gain characteristics.

10.7   ET, CFR, AND DIGITAL PREDISTORTION

ET is often considered in the same context as CFR and DPD, particularly when designers are weighing up which techniques can be used to improve efficiency with high PAPR signals. Questions such as "Do I need ET if I am using DPD?" or "Do I still need DPD if I'm using ET?" often arise.

ET can certainly simplify DPD or in some cases eliminate the need for DPD altogether. Equally, ET and DPD can happily coexist to achieve the maximum efficiency and linearity.

10.7.1   ET AND CREST FACTOR REDUCTION

Digital CFR is widely used in infrastructure applications to create waveforms with manageable PAPRs. The "native" PAPRs of OFDM signals can often exceed 10 dB, and sizing the PA design for these occasional peaks would increase cost and reduce efficiency. Crest factor reduction reduces the PAPR of the signal allowing the PA to on average operate with less back off—and hence higher efficiency than would be possible without CFR. However, this efficiency increase comes at the expense of EVM and/or ACLR, although the more sophisticated CFR algorithms allow control over balance between EVM and ACLR degradation by controlling the frequency placement of the inevitable distortion products.

The simplest form of CFR is to rely on the analog soft compression characteristic of a fixed supply PA to attenuate the peaks of the waveform. Although engineers will often say "we don't do CFR," this is often followed by the vague statement "we operate the PA 1 dB into compression at the peaks." Although simple and free, this approach is inflexible, as PAs typically have a soft compression characteristic, which the designer has only limited ability to alter, resulting in the need to operate with high back off. A further problem is that a fixed supply PA's compression characteristic is typically poorly controlled over temperature and mismatch.

Digital CFR techniques are therefore used in preference, to give stable, repeatable compression characteristics, allowing more precise trade-offs of EVM or ACLR. Various techniques exist, from a simple digital hard-clip-and-filter, a programmable AM/AM table, or sophisticated time-domain impulse cancellation CFR.

If digital CFR of the source waveform is available in an ET system, then it should be used, although the impact of PAPR on efficiency is likely to be less dramatic with ET than for fixed supply or APT operation. It is also possible to use the ET shaping table to implement a soft clipping CFR by flattening off the top of the shaping table to effectively limit the peak power. Although this approach comes "for free" with most ET implementations, it does result in the PA going more heavily into compression at the waveform peaks, which may introduce additional AM/PM distortion.

10.7.2   ET AND DIGITAL PREDISTORTION

As already discussed, the ET shaping table can be optimized for maximum efficiency, maximum linearity, or somewhere in between. When combined with DPD, the ET shaping table can be optimized for best system efficiency, while relying on the DPD to linearize the PA.

A wide range of precorrection techniques fall under the umbrella classification of "DPD," with widely varying complexity. It is helpful to categorize DPD systems as either open- or closed-loop and as correcting memoryless or memory (time-dependent) terms.

The simplest form of DPD is the open-loop memoryless system, which performs a static pre-correction of AM:AM and AM:PM. In an ET system, the AM:AM component can already be controlled via the ET shaping table, so it is also possible to use a phase-only correction in an ET system. A difficulty of the open-loop approach is having certainty that it is "stable enough" with respect to numerous system variables (temperature, part-to-part variation, VSWR, etc.)—and typically this is only really known once a large quantity of data has been gathered, which is seldom the case in the early stages of a development.

The memoryless DPD can also be made adaptive or closed loop, where a mathematical model of the PA's distortion is maintained and used to precorrect the transmitted signal. The PA model is updated when needed, using an observation receiver to capture a short section of the PA output, via an RF coupler. The system must also capture the same section of the original digital TX signal fed to the PA input, time align the reference buffer with the measurement buffer, and perform a series of relatively complex mathematical operations to update the PA model—which can take several seconds of DSP processing power. The associated overall power "cost" of adaptation depends critically on the frequency of update—which could be anything from once (factory calibration)—to occasional (on power up, or to the track temperature)—to more frequent (e.g., triggered by a change in load VSWR).

Both open- and closed-loop DPDs can also be implemented with "memory" terms, that is, the precorrection applied to the TX signal depends not only on the instantaneous amplitude of the signal, but also the amplitude of one or more prior samples. This significantly increases the complexity of both the precorrection and the adaptation tasks and is not commonplace in handset implementations at the time of writing.

10.7.3   USING ET WITH DPD

An important benefit of using ET with DPD is that operating the PA in ET mode significantly improves its stability over temperature. When the PA is in compression, the small signal gain of the PA has little influence on the output power, which is determined primarily by the ET supply voltage.

This can eliminate the need to run closed-loop or adaptive DPD, since the PA response is much more predictable in ET mode, potentially allowing use of simple open-loop correction. It also allows the option of correcting only the AM:PM component, although this does constrain the ET swing range to be optimized for linearity rather than efficiency as previously discussed.

Figure 10.14 provides linearity measurements for an ET PA using only isogain ET, followed by adding memoryless phase-only correction, gain and phase correction,

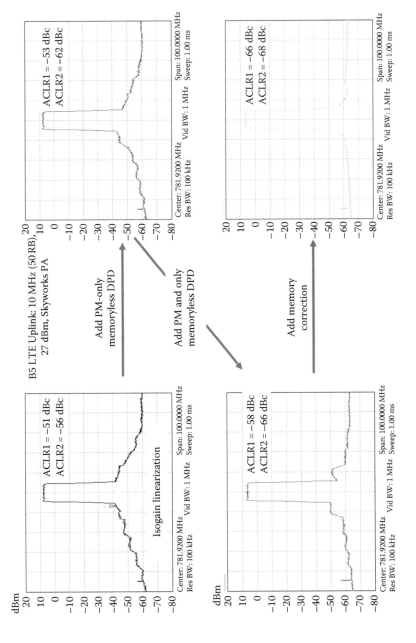

FIGURE 10.14   ET linearity with isogain, memoryless, and full-memory DPD.

and finally a full memory DPD solution. The ACLR progressively increases from −51 dBc with the isogain correction to −66 dBc with full memory DPD.

It is important to note that the PA's AM:AM and AM:PM response depends on the ET shaping table selected and can be extracted from the PA measurement surfaces as previously described. For each ET shaping table, the AM:AM and AM:PM responses could therefore be predicted and stored alongside the shaping table to configure an open-loop DPD.

DPD systems for fixed supply PAs require additional dynamic range in the IQ path DACs for "boosting" the peaks of the transmitted signal to compensate for the roll-off in PA gain. However, the gain of an efficiency optimized ET PA (i.e., one operating with "max efficiency" shaping) usually expands rather than reduces at peak power and hence requires slightly less dynamic range in the TX path. If the ET system is used to correct AM:AM (through the use of isogain shaping), and the DPD is used to correct only AM:PM, this may reduce the bandwidth requirements of the DPD, if only benign phase predistortion is necessary.

10.7.4 PRESPLIT AND POSTSPLIT DPD

When combining ET with DPD, it is possible to place the predistortion either before or after the ET path split. Performing the DPD at source, that is, before the ET path calculation, provides the simplest solution—the relationship between the PA supply voltage and PA input power is fixed, and the DPD linearizes the combined response.

However, DPD does expand the bandwidth of the TX signal quite significantly and also directly influences the ET swing range, which may cause performance limitations on the ET path. Implementing the DPD function after the ET/IQ split does not suffer from these disadvantages and also allows the DPD to correct for ET/RF path gain mismatch, if adaptive. However, in this location the DPD is only able to correct gain errors in the PA's low power region (where it is not in saturation) and so should be used in conjunction with isogain shaping in the PA's high power region.

10.7.5 ET AND MEMORY EFFECTS

Since ET is a time-domain process, a number of ET system degradations can give rise to memory effects similar, but not identical, to intrinsic PA device memory effects.

For example, a timing alignment error between ET and RF results in PA distortion which is dependent on the slope of the RF envelope, appearing as a widening of the instantaneous AM:AM and AM:PM plots.

Limited ET path bandwidth, for example due to inadequate ETIC frequency response, excessive PA decoupling or stray inductance in the ETIC—PA supply feed, also results in memory effect, which widens the instantaneous AM:AM and AM:PM plots.

Although some of these effects can be reduced using a memory DPD system, if available, not all are correctable (e.g., mistracking due to slew rate, switcher noise breakthrough), and care must be taken to reduce them at a system level.

10.8   ET INTO MISMATCHED LOADS

A potentially surprising by-product of ET is the significant improvement in linearity seen when operating into a load mismatch.

Although ET does not eliminate the need for antenna tuning, ET PAs do exhibit less variation in ACLR and EVM with VSWR than their fixed supply counterparts.

This effect is illustrated in Figure 10.15, which plots ACLR against PA output power for a PA operating in fixed supply and ET mode into a 3:1 VSWR at the PA output, representative of a 20:1 mismatch at the antenna plane. Measurements were made over 0°–360° in 45° steps. The input power to the PA was adjusted to maintain constant output power at the 1:1 VSWR value (27 dBm) for both ET and fixed supply operating modes.

It can be seen that the ET PA shows much less variation in ACLR than the fixed supply PA, and remains in specification regardless of phase angle, whereas the fixed supply PA is noncompliant at some phase angles.

This is because, unlike the fixed supply case where the PA is transitioning from linear operation to saturated operation, the ET PA is always in compression. The effect of the mismatch is to alter the gain and phase behavior of the PA—the average ET PA gain does vary—but this does not result in any dramatic transitions in the AM:AM or AM:PM curves.

Instead, the normally flat isogain response continues to "cut across" the fixed supply contours without introducing any catastrophic clipping (in contrast to fixed supply

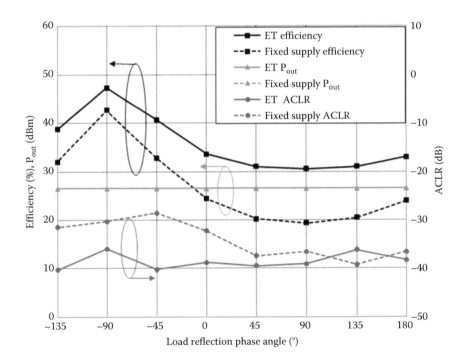

FIGURE 10.15   ET versus fixed supply performance with a 3:1 VSWR mismatch.

operation). The isogain does not remain perfectly flat, resulting in moderate AM:AM distortion. Figure 10.16 shows gain surfaces captured from an ET PA under 1:1 and 3:1 VSWR, with an isogain shaping table created from the 50 ohm surface (black line, left plot), and then applied to the mismatched surface (black line, right plot).

By observing the fixed supply curves in Figure 10.16, it can be seen that the onset of compression in fixed supply operation occurs 2–3 dB earlier under mismatch, causing the waveform to be severely clipped.

These results illustrate the improvement in linearity when an ET PA operates under mismatch at the output of the PA. However, in an LTE handset, mismatch often occurs at the output of the duplex filter which follows the PA. This does generate a more complex interaction with the ET PA, since the reflected power now arrives at the PA after being delayed through the filter. The delayed reflection can interact with the ET supply modulation, and adjustments to the ET/RF timing alignment may be found to improve performance under mismatch.

10.9 OPTIMIZING ET SYSTEM PERFORMANCE

10.9.1 ET CONFIGURATION MANAGEMENT

The ET-enabled handset must be preconfigured with ET shaping tables, ETIC settings, and timing alignment values at design time, together with the usual set of tables needed for APT voltages, PA bias settings, and settings for other RF front-end devices.

It is expected that at least one shaping table would be required for each LTE operating band, which can be optimized at design time. If DPD is used, an accompanying set of AM:AM and AM:PM tables can be generated.

Depending on the ETIC design and PA performance, additional performance gains may result from changing the ETIC settings (and possibly shaping table) based on average power level, channel bandwidth, waveform type/PAPR, or instantaneous RB allocation.

The designer must, therefore, develop a consistent methodology for generating and collating this data in a format that can be used by the layer 1 firmware.

10.9.2 PRODUCTION CALIBRATION

As previously discussed, ET introduces some new analog components into the RF signal path, namely the ET path DAC in the modem chipset, and the ET power supply modulator IC. ET performance is also influenced by the gain and delay of the RF path in the RFIC between the digital IQ signal and the input of the PA.

ET systems, therefore, require production-line calibration to remove several sources of potential unit-to-unit variation, namely

- ET path gain
- ET path offset
- ET/RF path gain errors
- ET/RF path timing alignment

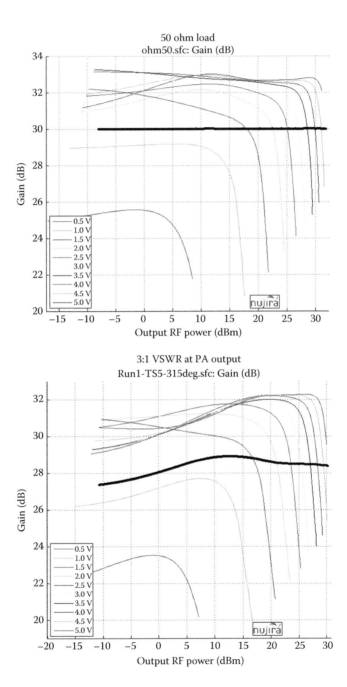

FIGURE 10.16 ET PA gain response under mismatch.

It is possible to devise algorithms for closed-loop calibration of these effects, which can be performed either one time on the production line, or alternatively used when the device is in the field for occasional or continuous self-calibration.

A measurement receiver in the modem chipset, similar to that used for adaptive DPD, can help significantly with these measurements, although it is also possible to run algorithms on external production-line test equipment to measure these impairments based on conducted or over-the-air RF measurements.

Real-time compensation may also be required for temperature effects based on design-time characterization of how parameters such as ET/RF timing alignment or RF path gain vary.

10.10 CONCLUSIONS

Although ET was initially introduced to reduce PA energy consumption in LTE handsets, the technology also delivers several other system level performance improvements, such as increased peak and average output power, improved linearity (even into mismatch), and better stability over temperature.

Almost all LTE chipsets now support ET, which requires the integration of a relatively simple digital signal processing block and a high-speed envelope DAC into the modem chipset. The ET path can be integrated into either the RFIC or the BBIC depending on the chipset partitioning. ET does need relatively high sample rates for the signal processing, with a good rule of thumb being 6× the RF channel bandwidth.

ETICs are now available from multiple vendors and usually include a combination of switch-mode power supply technology and high-bandwidth linear amplifiers. The interfaces between the modem chipset and ETIC have been standardized by the MIPI Alliance as eTrak and all ETICs deployed in terminals also use the MIPI RFFE low-speed serial control interface for control and configuration.

ET PAs require some basic modifications to support high-bandwidth supply modulation, such as removal of video decoupling components, but for optimum performance should be designed with ET in mind. For "ET-only" PAs, which do not need to support high-power APT operation, many traditional design goals (such as linearity) can be sacrificed to improve ET performance, requiring a different mindset from the PA designer. ET also unlocks the potential for CMOS PAs to be used with high-PAPR signals, such as LTE and LTE-Advanced.

As ET transforms the PA from a two-port analog component to a three-port software defined device, comprehensive system modeling is required to understand how performance is influenced by different system settings and component parameters. An accurate three-port model of the PA is needed, requiring new tools for lab measurement and PA characterization, which are now starting to appear from multiple vendors.

By intelligent design of the ET shaping table, the PA can in some cases be linearized without requiring any DPD, reducing development complexity and bandwidth requirements in the modem chipset. If ET is combined with DPD, the DPD complexity is significantly reduced, as the PA's distortion characteristics in ET mode are much more tightly controlled over temperature and process, allowing use of open-loop DPD with static gain and phase tables developed at design time.

ET does introduce some new complexity into the modem and RF system and has some impact on production calibration. But just like any other technology, once the industry overcomes the initial learning curve, ET will quickly become ubiquitous technology for 4G, 5G, and beyond.

Although ET has been around since the 1930s, it offers significant promise for the twenty-first century, extending the boundary of "software defined radio" from the modem into the RF front end.

11 MEMS Switching in the Handset RF Front End

Igor Lalicevic

CONTENTS

11.1 OVERVIEW

The focus of this chapter is on radio frequency microelectromechanical system (RF MEMS) switching technology and the benefits that RF MEMS switches provide for Long Term Evolution (LTE) wireless technology.

RF MEMS benefits are observed through different aspects of mobile phone performance improvement. The history of wireless technology development will be discussed from the perspective of improving the speed of data.

All of the major components in a typical mobile phone RF front end (RFFE) will be considered, and we will focus on their impact on performance with respect to the insertion loss, isolation, and linearity of the system.

Special attention is given to the next generation of RF switching requirements for the higher end of the LTE performance scale, and a particular emphasis is placed on the need for high-performance, high-quality, and highly reliable RF MEMS switches. This chapter includes a description of DelfMEMS RF MEMS switch technology as a solution for the next generation of switching and how it addresses the need for high performance while minimizing insertion loss, increasing isolation, and providing superior linearity performance in the front end.

11.2 DATA SPEED IMPROVEMENT AND WIRELESS TECHNOLOGY DEVELOPMENT

The headline findings of Cisco's latest Visual Networking Index states that by the year 2018, the composition of Internet Protocol (IP) traffic will shift dramatically and for the first time in the history of the Internet, mobile and portable devices will generate more than half of the global IP traffic. This is an excellent way to describe the history of wireless technology development, which has witnessed tremendous evolution and change in cellular standards over the past 35 years.

First-generation wireless telephone technology, also called 1G, was launched in Japan by NTT (Nippon Telegraph and Telephone) in 1979. 1G technology used analog transmission techniques for transmitting voice signals. This voice-only standard used a frequency-division multiple access (FDMA) technique where voice calls were modulated to a higher frequency of approximately 150 MHz and greater as it is transmitted between radio towers.

1G issues such as low capacity, unreliable hand-off, poor voice links, and complete lack of security, resulted in development of early digital systems, which became known as 2G: *second-generation* wireless telephone technology. These systems became available in the 1990s and used digital multiple access techniques, such as time division multiple access (TDMA) and spread-spectrum code division multiple access (CDMA), which enabled the first low bit rate data services.

2G systems offered higher spectral efficiency, the first data services, advanced roaming, and for the first time, a single unified standard was provided: global system for mobile communications or GSM. First employed in Europe in 1991, today GSM is still utilized across the world.

As the requirements for sending data over the air interface increased, general packet radio service (GPRS) and wireless application protocol (WAP) technologies were added to existing GSM systems. This was considered an advancement from 2G and became known as 2.5G technology, where packet-switching wireless application protocols enabled wireless access to the Internet.

Even though 2G supported data over the voice path, data speeds were typically a low 9.6 Kb/s or 14.4 Kb/s, which effectively made 2G a voice-centric system. Data speed improvement that 2.5G technology introduced with more advanced coding methods provided theoretical data rates of up to 384 kbps. However, limitations of packet transfer technology, which behaves in a similar way to a circuit switch called over the air, combined with low system efficiency and non-standardized networks across the world led to the birth of the 3G standard.

The initial planning for *third-generation* or 3G standards started in the 1980s, and the objective was to enable multimedia applications such as videoconferencing for mobile phones. However, over the years, 3G's focus moved toward personal wireless Internet access and the need to have connected access worldwide while achieving increased system and network capacity.

In the year 2001, the first commercial 3G network based on W-CDMA technology was launched in Japan by NTT DoCoMo.

The need to create a globally applicable mobile phone system specification resulted in the 3rd Generation Partnership Project (3GPP), which united seven telecommunications standard development organizations from Asia, Europe, and North America known as the "Organizational Partners." 3GPP has become the focal point for 3G and mobile systems beyond and it provided the requirements for 3G data speed specifications of up to 2 Mbps for stationary users.

To additionally enhance Universal Mobile Telecommunications System (UMTS)-based 3G networks, and to enable higher data speeds, features like High Speed Packet

Access (HSPA) have been implemented to provide data transmission capabilities delivering speeds up to 14.4 Mbps on the downlink and 5.8 Mbps on the uplink.

High Speed Packet Access Plus (HSPA+) or Evolved HSPA access was a further evolution of HSPA that became capable of delivering theoretical peak data speeds of 168 Mbit/s (downlink) and 22 Mbit/s (uplink). HSPA+ provided a migration method toward the next wireless standard data speeds without actually deploying a new radio interface.

The latest *fourth-generation Long Term Evolution (LTE) standard* uses a new air interface based on orthogonal frequency-division multiple access (OFDMA) digital modulation method. Long and diverse evolution paths of cellular standards toward a unique LTE solution are shown in the Figure 11.1.

The history of wireless technology development can be summarized as a history of data speed progression. Figure 11.2 summarizes the effects of the channel bandwidth and modulation scheme on the wireless standard's data speed.

The relationship between channel bandwidth and its associated data speed is evident and a crucial part of the direction of LTE technology development. The ever-growing need for mobile data created a tremendous need for more frequency spectrum. Over the last few years, we have witnessed an explosion of growth in the number of allocated frequency bands.

From the first 2G digital cellular standard that introduced quad-band system solutions, we have seen the trend of increasing the number of frequency band allocations with each generation. 3G typically supports up to 8 frequency bands in order to support global roaming and higher data speed needs. With 4G Long Term Evolution Advanced (LTE-A), the increase of allocated frequency bands is even more dramatic. Currently, over 40 bands are allocated to Long Term Evolution Frequency Division Duplex (LTE FDD) and Long Term Evolution Time Division Duplex (LTE TDD) applications.

Together with frequency band expansion, we have experienced significant data speed and capacity increases. Beginning with 2G at 14.4 kbps for both downlink

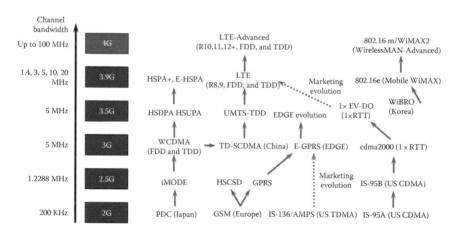

FIGURE 11.1 Cellular standards evolution.

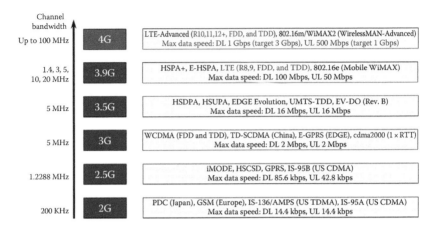

FIGURE 11.2   Data speed evolution.

(DL) and uplink (UL), currently LTE Cat9 will provide up to 450 Mbps downlink (DL) and 50 Mbps uplink (UL) data rates. But the market demand continues to grow. Even though the peak data rate targets for LTE-Advanced of 1 Gbps for the downlink and 500 Mbps for the uplink connection have not been achieved, revised targets to double or even triple these values are emerging.

Any solutions leading to a dramatic data rate increase will require more frequency spectrum. 3.5 GHz LTE bands 42 and 43 are a great illustration of this, together with the target to introduce 100 MHz of bandwidth for downlink carrier aggregation. Carrier aggregation is one key enabler for LTE-Advanced to meet the International Mobile Telecommunications–Advanced (IMT-Advanced) requirements of achieving peak data rates by a bandwidth increase methodology. Carrier aggregation or channel aggregation enables multiple LTE carriers to be used together to provide a bandwidth increase.

The LTE cellular standard is divided into categories, each of which caters to different requirements, as shown in Table 11.1.

TABLE 11.1
LTE Cellular Standard Categories

Categories		LTE Cat1	LTE Cat2	LTE Cat3	LTE Cat4	LTE Cat6	LTE Cat9
Data Rates	DL	10	50	100	150	300	450
(Mbps)	UL	5	25	50	50	50	50
Channel	DL	20	20	20	20 (CA)	40 (CA)	60 (CA)
Bandwidth	UL	20	20	20	20	20	20
(MHz)							
DL MIMO		Not sup.	2×2	2×2	2×2	2×2	2×2
Configuration							

Since the first LTE carrier aggregation solution provided LTE Cat4 that used 10 + 10 MHz aggregation, LTE Cat6 20 + 20 MHz carrier aggregation has delivered an additional 20 MHz of channel bandwidth that resulted in doubling data rates from 150 to 300 Mbps. The state of the art LTE-Advanced solution today, LTE Cat9, is using 60 MHz (20 + 20 + 20 MHz), three bands carrier aggregation in a 2×2 MIMO mobile handset configuration, with the capability to offer a 450 Mbps peak downlink data rate. While downlink data rates from LTE Cat1 to LTE Cat9 has increased 45 times, improving from 10 to 450 Mbps, uplink data rate speeds have increased only 10 times, improving from 5 Mbps for LTE Cat1 to 50 Mbps for LTE Cat9. It is noticeable that uplink data speed has stagnated since LTE Cat3.

11.3   RF FRONT END: INTRODUCTION

The large numbers of bands due to global roaming needs, carrier aggregation and MIMO have made RF front-end (RFFE) architectures for high-end smartphones an exceptionally complex and key bottleneck in achieving the market needed RF performance. Furthermore, due to economic pressures original equipment manufacturers (OEM) are looking to minimize their model variants. For some of them, the trend is to build one global phone model, which would by default assume the RF is capable of supporting 20 or even more bands. The list of needed RFFE components has become increasingly long, and all of them will need to meet required specifications in the "high number of bands" environment.

An example of a possible state of the art RFFE is shown in Figure 11.3 and it consists of main diversity antennas needed to achieve 2×2 Rx multiple-input multiple-output (MIMO), an antenna tuner and couplers, a front-end module (FEMiD) that includes antenna switches and filters, a diversity front-end module (divFEM) that includes antenna switches and Rx filters to accomplish 2×2 downlink MIMO and Rx diversity configuration. It further includes a multimode, multiband power amplifier (PA) module, a power management unit including an envelope tracking modulator, a radio frequency integrated circuit (RFIC) transceiver and multiple additional add-on components such as standalone filters, power amplifiers, switches, and so forth.

This complex RF environment introduces numerous challenges for selected RF components. Power loss or insertion loss, isolation, and linearity degradation are visible consequences of RF component's performance limitations and once again become a serious challenge and the main consideration in component selection.

An excellent example of the component performance impact on the end user is mobile phone battery life. According to the "top 10 smartphone purchase drivers" user study, battery life is the most important feature in the smartphone for over 50% of users. High-RFFE insertion loss (IL) has a direct negative impact on smartphone battery life and call quality.

Previously, the consumer 2G phone experience assumed that mobile handsets are typically used without recharging the battery for days. The explanation is fairly simple. 2G uses the Gaussian minimum shift keying (GMSK) modulation scheme. In this "constant envelope" method, the bit value does not affect the emitted signal's amplitude and its power level is constant during the burst. Lack of amplitude

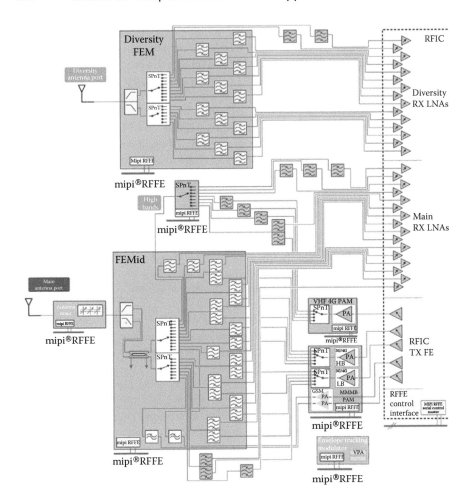

FIGURE 11.3 Possible LTE RFFE architecture.

modulation allows for the usage of efficient, saturated power amplifiers (PA) and thereby saves on current consumption; a critical issue for battery powered equipment. In this low-data rate modulation technique, low-RFFE IL of less than 1 dB had minimal impact on the RF current, of which the vast majority is spent on highly efficient PAs.

The situation became progressively worse with 3G standards. Battery life was visibly shorter compared to 2G phones, but for an average user everyday recharging was still not necessary. The 3G standard provided higher data speed by way of more efficient modulation schemes, which was an acceptable trade-off for decreased usability.

Binary phase shift keying (BPSK) is a major modulation technique within Universal Mobile Telecommunications Systems (UMTS) technology, and it requires PAs working in the linear operating mode. PAs are not operating in the efficient saturated region due to the high peak to average power ratio of HSDPA/HSUPA

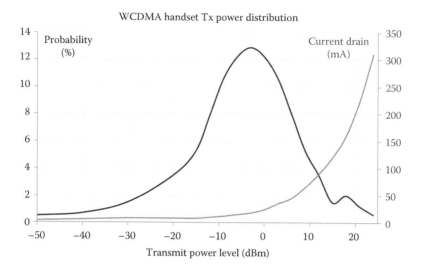

FIGURE 11.4 Transmitter current drain and 3G handset power distribution probability versus handset output power.

modulations that are applied in 3G mobile systems. PA efficiency, which defines battery supply consumption and PA adjacent channel power ratio linearity performance, works directly against each other. Therefore, linear PAs are, by default, less efficient from a current drain point of view. An interesting aspect of the 3G phone's battery life is presented in Figure 11.4.

The grey curve is an example of the PA current drain versus the mobile phone transmitted power. We see the exponential current drain increase versus the transmitted output power level. It follows then, that at a very high or maximum output power level, with maximum current drain consumption, our battery life will be significantly reduced.

From the black curve, which represents a 3G user's transmit power distribution, we see the probability of 3G phones operating at the highest power levels with high current consumption is below 2%. In conclusion, we can state that 3G phones have limited usage at the highest power levels where current drain is high, and the normal operating point for 3G phones is at lower power levels, which allow longer battery life.

A simple calculation using Figure 11.4 shows that an output power level of around 0 dBm will have current drain that is around 1/6 of the maximum transmitted output power current drain level. This will make a 3G phone battery life around 80% longer at the 0 dBm transmitted power level, compared to the battery life at the +23 dBm transmitted power level.

Any positive experience with long battery life for the 2G and 3G standards disappeared with fourth-generation phones. 4G mobile phones, LTE, and LTE-Advanced, have to be recharged daily and in a typical user case even more than once per day. The orthogonal frequency division multiplex (OFDM) modulation scheme requires PAs that use the same or a very similar design as 3G PAs, and the same linear PAs today are being used for both 3G and 4G applications.

From the battery life point of view, the crucial difference is that LTE networks typically require short bursts of high RF output power levels for data transmission. Going back to Figure 11.4, this means 4G phones are almost always used at maximum (+23 dBm) output power level that implies maximum current drain for 4G applications and a shorter battery life.

LTE-A with carrier aggregation and global roaming requirements are making things even worse. The necessity for LTE-Advanced to introduce additional LTE bands resulted in tremendous RFFE power losses. The typical RFFE IL for 2G was around 0.5 dB, and with the 3G standard this loss increased to around 1.5 dB. The IL for the 4G RFFE due to additional filtering needs and complex switching requirements can easily reach over 4 dB, which translates into over 60% of the power being lost. To offset and cover increased power losses and to meet the maximum antenna radiated power requirements, 4G PA's have to be capable of providing additional power that makes 4G PAs operate even less efficiently.

11.4   RF MEMS SWITCHES IN THE RF FRONT END

Extensive work has already been performed on RF MEMS technology solutions and it has been demonstrated that there can be a better antenna, filter, and power amplifier design by implementing RF MEMS. In addition to applications such as tunable antennas and filters, RF MEMS are also regarded as an ideal choice to implement RF switching.

Linearity, IL and isolation are requirements that have been defining transmit/receive and band select switching technology choices since the early days of wireless telephone technology. These performance requirements led to the selection of GaAs pseudomorphic HEMT (pHEMT) as the preferred choice for 2G switching technology. Increased 3G linearity requirements and a need for a high level of integration introduced solid-state silicon-on-sapphire (SOS) and silicon-on-insulator (SOI) technologies as a better alternative to GaAs pHEMT for RF switch applications. Recent advances in high-resistivity SOI technology has resulted in complementary metal-oxide semiconductor (CMOS) thin film SOI becoming the mainstream switching solution for 4G mobile phones because of its cost and performance benefits.

Very complex and "lossy" RFFE by default imply and emphasize a need for a high performing RF component that can bring crucial benefits and simplification to existing architectures. Additional complications are introduced with the inter-band carrier aggregation requirement, which requires the use of multiple active Tx/Rx paths within a single RFFE. This adds the usual impact on cost, performance and power, which are the most evident on the RF switch.

11.5   RF MEMS SWITCH BENEFITS

Multiple complexities result from the requirements to reduce intermodulation and cross modulation from the two or more receiver and transmitter paths. In this environment, for all RF components and particularly for the RF antenna switch, linearity performance is becoming a crucial specification.

The inherent high linearity of a MEMS switch is probably the most obvious benefit of RF MEMS switching because it can enable uplink carrier aggregation. Equally as important, especially at the higher frequency operating bands, RF MEMS ultra-low IL is seen as an ideal solution to maintain high transmitter power efficiency and a low-receiver noise figure. MEMS inherent design characteristics provide the high isolation performance needed between the transmitter and receiver.

The 3GPP standard is used by the industry to determine the degree of linearity required to avoid interference with other devices on the network. This is done by specifying the third-order input intercept point (IIP3).

Each new generation of cellular network required progressively higher linearity. In Table 11.2, we can see that the 2G requirement for switch linearity was an IIP3 of 55 dBm, the 3G switch requirement was 65 dBm, the LTE switch IIP3 requirement is 72 dBm, and the LTE-Advanced antenna switch with up-link carrier aggregation capabilities will have to meet an IIP3 of 90 dB. Currently, the dominant solid-state switch technologies such as SOI or SOS are not capable of achieving an IIP3 target of 90 dBm.

It is essential to start thinking about the existing switching technology limitations, which will have to be addressed with alternative solutions or a completely new technology that can meet existing challenges and provide the needed performance.

RF MEMS technology provides answers for these LTE-Advanced challenges with the benefits of providing better call quality, longer battery life, and enabling the 90 dBm requirement for uplink carrier aggregation to improve smartphone performance.

In order to compare the performance of the different switch technologies, an industry standard figure of merit (FOM) has been established, which is defined as $R_{ON}C_{OFF}$, and is expressed in femtoseconds (fs).

R_{ON}: ON Resistance (unit: Ω mm)
C_{OFF}: OFF Capacitance (unit: F/mm)

In Figure 11.5, DelfMEMS RF MEMS switching technology is compared to solid-state technologies. Its outstanding figure of merit provides a 10× improvement over the next best solution.

TABLE 11.2

Network Linearity Requirements

Network	Linearity (IIP3) (dBm)
2G	55
3G	65
3.9G	72
4G (UL CA)	≈90

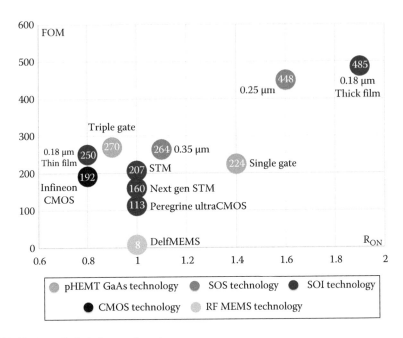

FIGURE 11.5   Switch figure of merit.

Solid-state switch designs are heavily dependent on the field effect transistor's (FET) channel resistance when it is ON (R_{ON}), and the switch's isolation cannot be better than the FET's total capacitance when the switch is in OFF state (C_{OFF}). Therefore, it follows that for high throw-count solid-state switches design, OFF state switch branches will become a dominant source of non-linearity, IL and isolation degradation. Equally important are degradations that solid-state technology will experience at higher frequency operation.

By contrast, DelfMEMS RF MEMS switches retain ultra-low IL (0.35 dB @ 3.5 GHz), very high port-to-port isolation (35 dB at 3.5 GHz) and ultra-high linearity (85 dBm up to 98 dBm) even for high multi-throw switch configurations and at higher frequency operation.

These parameters make RF MEMS switches a perfect candidate for the 4G multiband environment where RF components have to demonstrate minimum power loss degradation at frequencies of up to and above 3.5 GHz. Figure 11.6 compares DelfMEMS versus solid-state switch technologies in the SP12T RF switch configuration.

The power loss reduction that can be realized replacing existing the solid-state switch with the DelfMEMS switch leads directly to a significant battery power savings and increased receiver sensitivity. These improvements become even greater at the new 3.5 GHz LTE bands, emphasizing RF MEMS broadband design and its ability to maintain superior performance with increasing frequency.

The block diagram presented in Figure 11.7 assesses the potential power savings when comparing the DelfMEMS RF MEMS switch to existing solid-state antenna

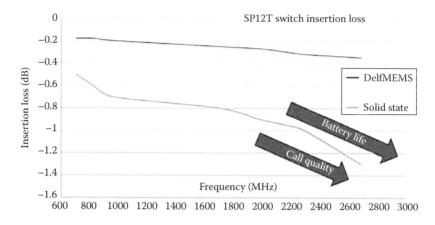

FIGURE 11.6 SP12T switch insertion loss.

switches. Here, transmission current drain is the dominant contributor to smartphone battery life, and current savings percentages across bands have been considered as equivalent to percentages of longer talk time.

IL of the existing solid-state switch solution was compared with DelfMEMS RF MEMS switching solution in decibels. A simple calculation for different bands was performed where current drain loss (CDL) is calculated and expressed in comparative percentages when solid-state switches are replaced by ultra-low-loss RF MEMS switch. IL reduction due to the replacement of the "lossy" SOI switch will result in up to *17% longer talk time*. The same logic applies on LTE-Advanced call quality and data reception quality.

As stated earlier, global roaming and world phone requirements for smartphones are creating a high-power loss RFFE environment that is particularly effecting both diversity and main receiver RF paths.

Receiver sensitivity is one of the key specifications for any radio receiver. Considering that call quality is equal to Rx sensitivity, any increased RFFE IL degrades Rx sensitivity by the same amount and therefore, Rx sensitivity is degraded directly impacting phone call and data reception quality.

The additional block diagram presented in Figure 11.8 assesses Rx sensitivity improvements comparing DelfMEMS RF MEMS switches to existing solid-state switches for both main and diversity antenna switch paths. Decreased RFFE IL in decibels will improve Rx Sensitivity by exactly the same amount.

The simple calculation where IL is max power loss for the switch expressed in decibels, and RxSI is calculated receiver sensitivity improvement expressed in decibels, demonstrates up to 1.1 dB Rx sensitivity improvement, which is equivalent to an impressive 29% Rx sensitivity improvement.

In conclusion, the LTE-Advanced market that is driven by higher downlink and uplink data speed requirements comes with a price of even less battery life and degraded signal reception quality. RF MEMS switches as a broadband design for high-frequency bands, offers the simple and effective solution to reduction of RF

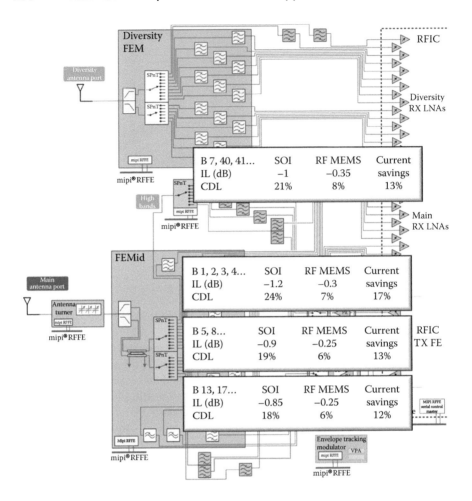

FIGURE 11.7   Possible LTE RFFE architecture—talk time.

FE losses while providing exceptional band-to-band and TX/RX isolation and the needed high linearity to enable uplink carrier aggregation.

11.6   DelfMEMS RF MEMS SWITCH SOLUTION

The typical RF MEMS switch structure uses either a cantilever beam or a bridge and features highly conductive electrodes, which are electrostatically actuated in order to create an ohmic contact on a conducting line. The result is mechanical switching. These typical basic structures carry several serious issues such as stress on the anchors, a tendency for stiction, low-switching speed, and metallic creep in the beam.

DelfMEMS RF MEMS switch design offers an innovative approach to address these problems, instead of trying to reduce their effect. As a result, the switch

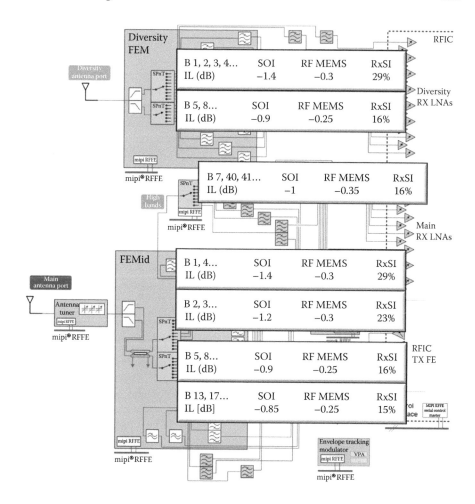

FIGURE 11.8 Possible LTE RFFE architecture—call quality.

simultaneously offers increased performance in terms of isolation and IL along with its unique design. This has been achieved through the development of a unique anchorless structure for mechanical RF switching.

DelfMEMS switch solution was from its very inception designed to address the question of RF MEMS reliability. It features a free moving flexible membrane, known as Free-Flex™, that carries the switch contact area and is held and positioned by two sets of pillars and stoppers. The membrane is always in a known controlled state as it is electrostatically actuated by two sets of electrodes. The electrostatically ON state is achieved by making physical contact between the membrane contact area and the transmission line and the similarly controlled OFF state is achieved by keeping a physical distance between membrane contact area and the transmission line. This means that the switch contact area will either be attracted to the conductive line or repelled from it.

Complete control of the MEMS membrane allows for an increased gap between contact area and transmission line in the OFF state, which is directly linked to the switch isolation and allows for switch resetting in the unlikely event of stiction.

Moving from the ON state to the OFF state is made through an electrostatic active actuation, which de-correlates between restoring forces, contact forces, and the membrane mechanical properties. In addition, the effects of mechanical bounce are substantially reduced. Importantly, this does not depend only on the elastic restoration forces. This advanced electrostatic actuation results in a very short switching time, typically less than 3 μs.

The ability to have a reduced gap between the membrane and the transmission line is a major advantage of the DelfMEMS switch structure. It ensures that an increased ON-state contact force is achieved with reduced actuation voltages and consequently delivers ultra-low insertion losses. Due to the reduced gap, the maximum deflection of the membrane will be reduced as well, and as a result, membrane mechanical stress and the creep effect decreases. The reduced gap combined with a unique metallurgy for the contacting surfaces results in a highly reliable RF MEMS switch.

Thanks to the DelfMEMS original design approach, RF MEMS switches can be used effectively for the first time as an RFFE switching solution.

11.7   CONCLUSIONS

LTE technology with its global roaming requirements, carrier aggregation, MIMO design approach and migration to higher frequency bands has made RFFE architectures for high-end smartphones exceptionally complex and a key bottleneck in achieving the RF performance levels required by the market.

Mobile handset battery life, call quality, and data throughput are the most obvious victims of this excessive complexity.

Extensive work has already been performed to improve IL, isolation, and linearity of existing RFFE systems, but the results remain largely disappointing and, when compared to 3G, no significant improvements have been demonstrated to date.

A number of companies are endeavoring to design tunable frequency component solutions that will reduce existing RFFE complexity. However, many of these approaches are fraught with difficulty and risk. DelfMEMS simple, yet innovative, approach to resolving the RFFE complexity problem is to replace specific switching components with a highly improved switching solution as has been discussed in this chapter, a solution that provides significant overall system performance benefits.

12 Case Study of Tunable Radio Architectures

Alpaslan Demir and Tanbir Haque

CONTENTS

12.1 INTRODUCTION

Recent user demand for ubiquitous, instantaneous over-the-air access to multimedia-rich content has forced related industry segments such as service providers, device manufacturers, technology developers, as well as regulatory bodies to collaborate more than ever before. In the United States, for example, the Federal Communications Commission (FCC) is trying to balance incumbent user demands while creating opportunities for others to trigger economic growth especially in TV white spaces (TVWS),[1] also known as digital dividend, the spectrum that is released as a result of television broadcasting companies switching from analog to digital-only platforms. The FCC's initiatives in TVWS were followed by countries such as Canada, Australia, Great Britain, and others. As part of stimulating economic growth, the FCC is working together with industry stakeholders and the Department of Defense (DoD) to enable utilization of the 3.5 GHz Citizens Broadband Radio Service (CBRS) band for multitier communications.[2] The FCC is also looking to facilitate mobile radio services in bands above 24 GHz to enable the fifth-generation (5G) mobile services within the context of broader efforts to develop technical standards.[3]

Increasing data traffic demands, especially due to social networking and related multimedia-rich content delivery, resulted in heavy data traffic burden for LTE technology over an all IP network. The issues impose a constantly expanding load on packet data distribution. As a result, data off-loading is defined as a key solution to ease core network congestion problem in the 3rd Generation Partnership Project (3GPP) standardization efforts.[4]

To catch up with user demands and application developers' limitless imagination, the device manufacturers are developing very powerful, yet complex and compact handset designs that enable multiple applications to run simultaneously. Also, the multiband, multimode enabled devices became a design requirement due to the fractional reuse of RF spectrum and relevant standardization efforts. However, due to the compact design constraints, enabling simultaneous operations of multiple radio access technologies (RATs) can cause in-device self-interference. The deployment of two separate RATs adjacent in space and spectrum may not be always possible. The operation of one RAT with high transmit power levels can desensitize another RAT's receiver in space-limited compact devices when the spectral separation between the two is limited.[5]

The tunable radio architectures may address a much broader solution spectrum for the problems mentioned earlier. However, in this chapter, a deployment scenario that is related to radio frequency front end (RFFE) in-device coexistence issues for simultaneous operations of LTE and Wi-Fi RATs is emphasized. Then, we define a system-level context aware RF front-end (CARF) approach to create a framework to enable intelligence in the RFFE domain. Also, an exemplary perspective on the protocol stack and overall system integration is captured. We conclude our study with some RF level treatment of the LTE and Wi-Fi coexistence issues followed by an exemplary control flow implementation that can enable adjacent and simultaneous operation of multiple RATs.

12.2   DEPLOYMENT SCENARIO

In this section, we focus on a deployment scenario to prepare the groundwork necessary to show how, in a system's perspective, some of the issues related to simultaneous operation of Wi-Fi and LTE RATs in multimode devices are solved. These issues can also be referred to as in-device, coexistence, or multihoming problems.[6]

In wireless network deployment a coverage area is confined to the maximum distance over which the network-related services are reasonably provided from the node (i.e., base station, Wi-Fi access point [AP]) that is connected to the service providing entity. The user demands may be served by a local network that may or may not have a gateway to the Internet. The wireless network coverage may also be defined as a function of path loss over distance to achieve a target signal-to-noise ratio (SNR). For example, a Wi-Fi network deployment may be designed to achieve minimum SNR target of +20 dB to define its wireless network coverage to guarantee an acceptable signal quality at network edge or anywhere the users are expected to appear. In other words, the minimum SNR at network edge is guaranteed to be more than +20 dB. Although the Wi-Fi systems are still operational below +20 dB SNR levels, the connectivity and data rates may become problematic at around +10 dB SNR. A Wi-Fi deployment with a single AP may operate at 24 dBm equivalent isotropic radiated power (EIRP) with less than +6 dBi omnidirectional antenna gain at 2.4 GHz. The effective noise power can be calculated by using Equation 12.1 at room temperature for bandwidth of 20 MHz as −101 dBm. If the receiver noise figure is assumed to be 7 dB, the effective noise power can be calculated as −94 dBm. Since we define the network edge SNR at +20 dB minimum, the target received power at the receiver antenna port must be at least −74 dBm (receiver noise power + SNR). Therefore, the path loss of 98 dB (Tx radiated power + receiver noise power − SNR at network edge) can be tolerated. By using Equation 12.2, the Indoor Hotspot for non–line of sight at 2.4 GHz as described in Annex B of,[7] the path loss of 104 dB corresponds to 66 m in distance. The coverage may be described as a circle with 66 m in radius where the AP is located at the center. The coverage we described is based on transmit power, path loss, and receiver characteristics; however, the coverage can also be defined based on service availability:

$$\text{Noise Power}_{\text{dBm}} = -174 + \text{Noise Figure} + 10\log_{10}(\text{BW}_{\text{hz}}) \tag{12.1}$$

$$\text{Path Loss}_{\text{dB}} = 43.3\log_{10}(d_m) + 11.5 + 20\log_{10}(\text{f}_{\text{c,GHz}}) \tag{12.2}$$

In a coverage area as shown in Figure 12.1, a macrocell tower is deployed to serve UEs via an LTE-based access link. In cellular networks, Macrocells provide radio service coverage via a high-power base station and are designed to provide coverage of up to 35 km. A home enhanced node B (HeNB) and a Wi-Fi AP are also deployed, as shown in Figure 12.1, in the same coverage area to enhance user experience and enable traffic off-load capability for the LTE core network. The HeNB is connected to the LTE core network with HeNB gateway (GW). In Figure 12.1, all UEs are located within LTE Macrocell coverage and capable of using LTE Macrocell access link for all services. Some of the UEs, as depicted in Figure 12.1, have additional options for the required services and can possibly utilize additional access links. The UEs highlighted in black in Figure 12.1, for example, are capable of accessing both HeNB and AP access links in addition to the LTE Macrocell access link. In some use cases, the LTE and the Wi-Fi RATs may be activated simultaneously to increase the data rates as a result of higher traffic demands based on application, user preferences, and network operator policies.

The network and regulatory policy related databases are also captured in Figure 12.1, where the policies under an operator's network can impact how a UE can access network services. The network policies may force the UEs with additional access capabilities to off-load their cellular traffic to Wi-Fi links. In addition, the regulatory policies may set rules governing how a modem behaves under certain operational conditions. For example, the regulatory policies may set limits to spectral emission masks based on a specific frequency band to limit interference to other devices as well as harmful levels of radiation to humans.

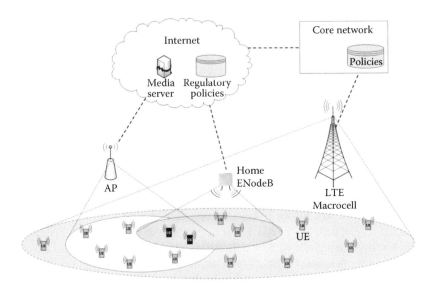

FIGURE 12.1 High-level system deployment scenario.

12.3 CONTEXT AWARE RFFE RESOURCE MANAGER WITHIN THE PROPOSED SYSTEM ARCHITECTURE

The Context Aware RFFE Resource Manager (CARF RM) is defined to create a framework to enable intelligence in RF domain based on system-level requirements by utilizing the modem's RFFE HW–related resources such as tunable antennas, tunable filters, adaptive waveforms, and analog and/or digital interference cancellation blocks. While improving spectrum utilization efficiency and battery power consumption, the defined system architecture and supporting interfaces enable a cross-layer scheme that optimizes user experience based on system parameters such as location and interference mapping, network and regulatory policies, user preferences, and QoS requirements.

In the case of inter- and intraband carrier aggregation, the CARF RM may adjust the PA efficiency and linearity based on the TX spectral mask requirements. The regulatory policies may impose different out-of-band spectral emission requirements on the operational band(s) as well as the channel(s). The CARF RM may also tune adjacent channel blocking (i.e., receiver selectivity) and the RFFE linearity (i.e., the dynamic range) to optimize battery power utilization while adhering to the QoS needs and the regulatory policies.

Figure 12.2 defines the high-level block diagram of CARF RM. The CARF RM collects system metrics, network and regulatory policies, user preferences, sensor fusion results, and any other inputs such as battery status, temperature, and runs algorithms to achieve a desired level of user experience. The CARF RM, then, sends the resulting commands to the RFFE controller. The RFFE controller makes adjustments on RFFE hardware resources based on the received commands.

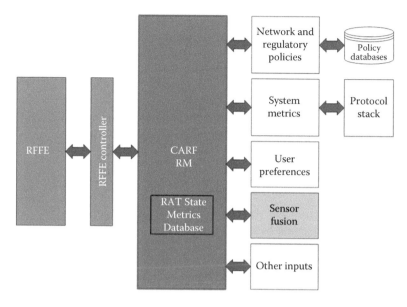

FIGURE 12.2   CARF RM system architecture with RAT state metrics database.

The RFFE controller can be flexible and can leverage additional performance settings to meet performance objectives with RFFE resources as they become available. For example, emerging device technologies can be incorporated to the RFFE controller that may provide new capabilities to further optimize the RFFE performance.

The RAT State Metrics Database, as depicted in Figure 12.2, includes certain system parameters such as channel assignment, Tx power limits, and RAT activity status. In the case of Multi-RAT operations CARF RM utilizes the RAT State Metrics to enable or disable self-interference cancellation methods. The details about how the CARF RM make use of these metrics will be given in Section 12.3.1.

In order to ground the CARF RM concept we define what each block entails in the following sections as shown in Figure 12.2.

12.3.1 SYSTEM METRICS

CARF RM requires protocol stack integration in order to extract system-level knowledge to adjust RFFE parameters. System metrics such as QoS indication on each packet or the group of packets and active channel SNIR levels give CARF RM necessary inputs to run certain algorithms and create commands to optimize RFFE HW resources. For example, the low QoS required marking on a packet can trigger the reduction in linearity requirements on the transceiver chain. Conversely, the high QoS marking on a packet can push the transceiver chain to be more linear. Adjustments in battery power management depends on the fact that the high linearity in RFFE requires more battery power consumption. The example of QoS-driven instant linearity scheme optimizes battery power consumption by allowing the transceiver chain to use just enough battery power for the amount of linearity to be maintained. The system metrics can also provide a sense of segregation such that application-driven CARF RM algorithms enables behavioral changes based on activity generated by a specific application. Enabling packet inspection may give CARF RM necessary information to create application level behavioral model at RFFE level. For example, certain applications and/or contexts such as emergency call may be set to receive higher performance at RFFE level and certain others not.

12.3.2   Sensor Fusion

The CARF RM heavily relies on sensory information to scan spectrum in order to create interference mapping based on location. It is assumed that the sensor fusion block is capable of location estimation via GPS or other means. The CARF RM obtains the location info from the sensor fusion block. It will set scanning requirements such as threshold, channel bandwidth, search span, and scanning method and then send commands to the sensor fusion block.

12.3.3   User Preferences

The CARF RM is capable of optimizing user experience based on user preferences. For example, the user can override RFFE performance criteria based on time duration and application. A user may choose to set RFFE in high-performance mode for an important phone call based on a contact's higher associated group set by the user by associating names to different RFFE performance levels while the penalty drains the battery power faster to receive a higher quality of experience. A user can also set predefined performance metrics to be adhered by RFFE resources for specific applications based on packet inspection that identifies activity for the application.

12.3.4   Network and Regulatory Policies

The CARF RM adds regulatory and network policies as constraints to solve optimization problems for the best possible utilization of the RFFE resource. For example, the transmit power and linearity can be controlled to create desired performance at RF level based on location, band selection, and regulatory policies relevant to spectral mask emissions. A network operator can choose to set additional policies for interference coordination for its own network.

12.3.5   RFFE Controller

This entity maps the received commands from the CARF RM to the Macro procedures to control RFFE HW resources such as setting the tunable filter effective bandwidth and amount of filtering required or enabling/disabling analog and/or digital self-interference cancellation blocks.

12.3.6 RFFE HARDWARE

The RFFE HW is comprised of spectrum and interference detector, circuit level sensors/detectors, and a transceiver chain that contains PA, LNA, tunable antennas and filters, baseband processing capabilities, and all other configurable elements. The RFFE HW is connected to the RFFE controller by means of Macro procedures that contains SW drivers for each RFFE resources. Each configurable element may be associated to a unique driver for operation and control.

12.4 PROTOCOL STACK AND OVERALL SYSTEM INTEGRATION

The purpose of this section is to facilitate a command-based messaging for the CARF control interfaces. Figure 12.3 shows a high-level block diagram of an exemplary CARF implementation. The CARF RM can send commands to the RFFE controller. The RFFE controller implements algorithms to interpret and initiate the desired RFFE functionality by controlling the RFFE resources. The messaging exchange is mainly at the interface between the CARF RM and the RFFE controller as illustrated in Figure 12.3. The CARF RM sends commands to the RFFE controller via this interface and messaging structure. The commands are processed by the RFFE controller that results in interpretations being sent to the RFFE Macro procedures. The RFFE Macros process the commands and controls the parameters in the RFFE resource elements.

The CARF RM entity that gives intelligence to the RFFE may be implemented on both sides of the RFFE interface to accommodate low- and high-speed messaging very efficiently as illustrated in Figure 12.3. Some of the low-level messaging may be handled directly at the RFFE controller level to enable very quick respond times.

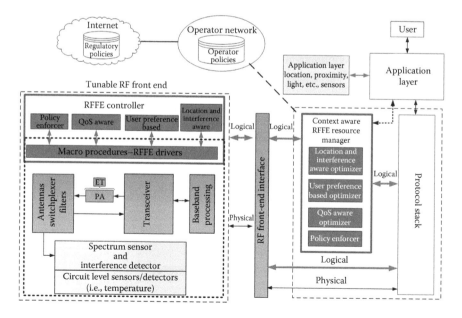

FIGURE 12.3 Context-aware RFFE protocol stack integration.

The messaging described here can be overlaid onto any existing or future standard or custom, semi-custom, or proprietary interfaces such as SPI, for example. These messages specify the control interfaces between the CARF RM and Macro procedures (i.e., RFFE drivers) via the RFFE controller.

By structuring messages on the interfaces based on CARF concepts described so far in this chapter, the benefits such as control latency, battery power efficiency, and radio faults circumvention can be realized. In addition, the learning mode support may enhance performance and battery power consumption.

As part of context awareness, social setting can also be supported. The user, for example, may associate higher or lower RFFE performance to a particular contact in a specific social media platform such as Skype, WhatsApp, Viber, and LINE contacts.

The CARF control requires an efficient and capable real-time interface that exists between the CARF RM and the RFFE controller as shown in Figure 12.3. The CARF RM can also affect the interfaces between the RFFE controller and the RFFE resource such as tunable antennas and tunable filters. The interface must support control of a diverse collection of RFFE resource elements. A myriad of control parameter formats require the interface to be highly flexible. The problems confronting the architecture of a suitable interface include, but are not limited to, issues that are described here.

The interface should be low latency for real-time control of RFFE resource elements. For example, some controls are only valid for a short time due to fast changing context that makes low latency essential. Other instances requiring low latency are also possible.

The interface should be low power to conserve battery life and the minimization of message traffic could reduce power. The programming and development of applications should be simplified. Initiating functional modes with abstracted or combined controls can help to simplify application development.

The real-time debugging should be enabled by the interface. This is desirable to fix problems and tune performance. The operation during fault conditions and fault circumvention should be supported where possible.

The emergency communications should be supported. For emergency context, the RFFE performance should be highest possible to increase the likelihood of successful communication. The interface should support social settings chosen by the user or service provider. For example, the user may require utmost performance for a mission critical phone call or launching a particular application.

The operation during low battery power conditions should be supported. The communication bandwidth can be reduced when battery power is at a critical low level. Other optimizations besides limiting bandwidth are also possible.

The learning should be enabled to facilitate optimization of performance and battery life. Other benefits from learning are also possible.

The interface should support complete control of all RFFE operating parameters. It should not limit functionality, operation, or performance.

12.4.1 MACRO PROCEDURES AND **RFFE HW DRIVERS**

The block called "Macro Procedures–RFFE DRIVERS" as shown in Figure 12.3 interfaces the RFFE controller to the transceiver chain and a reconfigurable scanner. It simply runs controlling SW interface to the RFFE HW resources. For example, it may map the command received at the RFFE controller to set high RFFE performance into Macro procedures that may do several adjustments such as increasing biasing current and voltage, changing PLL loop filter bandwidth, and so on.

12.4.2   CARF RM Optimizer

The interfaces within CARF RM enable all the decision-making-related information among location and interference aware optimizer, user preference–based optimizer, QoS-based optimizer, and policy enforcer.

12.4.3   Location and Interference Awareness

The main functionality of this entity is to command and collect interference mapping of the communication equipment based on location. It also runs algorithms to determine how much interference suppression is required in order to have desired SNIR for the baseband data demodulation.

The entity may employ closed- or open-loop interference suppression algorithms by commanding the RFFE controller that sets the desired parameters for tunable antennas and/or tunable filters and enables or disables analog and/or digital interference cancellations blocks.

12.4.4   User Preference

This entity receives user preferences and provides information to other controlling blocks under CARF RM. It sends commands to the RFFE controller. For example, the user may prefer for a period time that the RFFE performance is to be very high or very low. Upon the reception of user preference, the user request is translated as a command and sent to the RFFE controller by the user preference–based optimizer. The user may prefer that a particular application is associated to a RFFE performance level.

12.4.5   QoS Awareness

The main functionality of this entity is to observe protocol stack–based system metrics and optimize RFFE performance. For example, in the case of IEEE 802.11 protocol stack, when communication is done packet by packet, the QoS aware optimizer observes the SNIR level required for each packet and sends commands to the RFFE controller in order to prepare the RFFE for each packet transmission.

12.4.6   Policy Enforcer

This entity collects the policies that govern each decision made within CARF RM. For example, policy may set max transmit power for a given location. This information is made available to all other deciding entities and sent to the RFFE controller to make sure that the PA output power is not exceeding the limit.

12.5   RF PERSPECTIVE ON SIMULTANEOUS OPERATION OF LTE AND Wi-Fi RATS

An increasing number of smartphones and other types of mobile communications devices today are equipped with multiple radio transceivers in order to facilitate the ubiquitous access to various networks and services. As an example, consider a UE equipped with LTE and Wi-Fi transceivers as shown in Figure 12.4. One outcome of this sort of hardware deployment is coexistence interference between collocated radio transceivers.[5,8,9] We examine the coexistence-related self-interference problem and explore possible system-level solutions that illustrate the utility of the CARF-RM.

It is conceivable that both the LTE and Wi-Fi transceivers are physically located at their respective cell edges. The power of the LTE uplink signal, for example, may be much higher than the Wi-Fi received signal. Due to the close proximity of the two transceivers, the LTE (aggressor) transmit signal will desensitize the Wi-Fi (victim) receiver. The deleterious effect of this sort of in-device self-interference can be mitigated through the employment of high-performance bandpass filters and careful frequency planning (e.g., carrier frequency separation). However, for certain deployment scenarios like that shown in Figure 12.5 where LTE and Wi-Fi transceivers operate in adjacent frequency bands, the only LTE and Wi-Fi carrier frequencies available at a particular physical location and time may be located at the high and low edges of the LTE and ISM bands, respectively. In such scenarios, the current state-of-art filter technology may not provide sufficient isolation between the two transceivers. The average out-of-band rejection of current state-of-art FBAR bandpass filters[9] for the LTE and ISM bands is summarized in Table 12.1.

When the LTE transmitter is active, the self-interference seen by the Wi-Fi receiver may be classified into three categories. These are the out-of-band (OOB) blocker, nonlinear interference, and noise. The OOB blocker consists of the leaked and reflected versions of the LTE transmit signal itself. The nonlinear interference consists of the intermodulation distortion terms, for example, in the transmit signal's adjacent and alternate channels created by the LTE transmitter. The rejection of the ISM bandpass filter in the adjacent LTE band determines the OOB blocker level seen by the Wi-Fi receiver when the collocated LTE transmitter is active. On the other

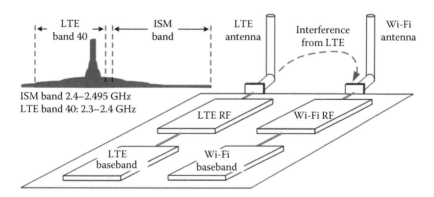

FIGURE 12.4   Coexistence interference within the same UE.

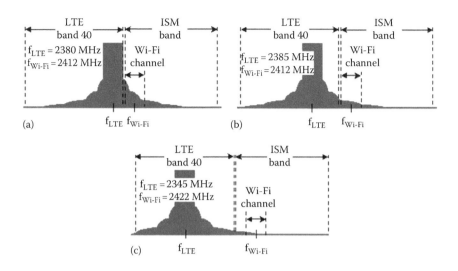

FIGURE 12.5 LTE band 40 and ISM band interference scenarios: (a) poor, (b) moderate, and (c) high spectral separation.

TABLE 12.1
Average Out-of-Band Rejection of LTE and ISM Bandpass Filters

	Frequency (MHz)				
	2300–2380	2380–2400	2402–2414	2415–2418	2419–2480
ISM BPF (ξ_{Wi-Fi}), dB	>39 dB	4 dB	—	—	—
LTE B40 BPF (ξ_{LTE}), dB	—	—	10 dB	>35 dB	50 dB

hand, the rejection of the LTE bandpass filter in the adjacent ISM band determines the level of LTE transmitter noise and nonlinear distortion that may fall in the ISM band.

The effect of OOB blockers and spurious emission interference (nonlinear distortion and noise) is studied in.[9] The total impact of blocking and spurious emission is considered. The results are summarized in Table 12.2. The table entries are the level of Wi-Fi (victim) receiver desensitization in dB. Desensitization is defined as $10 \log_{10}(\alpha)$ where α is the total coexistence interference to the receiver thermal noise floor ratio. For a given entry in the table, the column represent the aggressor (LTE) transmit frequency and the row represents the victim receive frequency. It is assumed that the LTE transmitter output is set to its maximum level (+23 dBm) and the Wi-Fi receiver is set to its maximum gain setting (i.e., operates at maximum sensitivity). The victim receiver desensitization is categorized into three levels—level 1 (>50 dB), level 2 (>10 dB and <50 dB) and level 3 (<10 dB). In the most extreme case,

TABLE 12.2

Coexistence Interference (Blocking and Spurious Emission) Impact from LTE in Band 40 on Wi-Fi in ISM Band

Wi-Fi (MHz)	LTE (MHz)									
	2310	2315	2325	2335	2345	2355	2365	2375	2385	2390
2412	>10 dB / <50 dB	>10 dB / <50 dB	>10 dB	>10 dB	>10 dB	>10 dB	>10 dB	>50 dB	>50 dB	>50 dB
2422			<10 dB	<10 dB	<10 dB	>10 dB / <50 dB	>10 dB / <50 dB	>50 dB	>50 dB	>50 dB
2432								>50 dB	>50 dB	>50 dB
2442								>50 dB	>50 dB	>50 dB
2452, 2462, 2472								>10 dB / <50 dB	>50 dB	>50 dB

where the aggressor and victim are in close spectral proximity, the victim receiver is desensitized by more than 50 dB. Note that the impact of the high OOB blocker on the Wi-Fi receiver LNA (low-noise amplifier) is ignored in.[9] In typical applications the LNA gain might be reduced to prevent it from saturating. This, in turn, will desensitize the receiver. Alternatively, the OOB blocker could be removed or canceled at the LNA input.

The OOB blocker and the in-band self-interference together reduce the effective sensitivity level of the Wi-Fi receiver. As an example, consider the three scenarios depicted in Figure 12.5. When the LTE transmitter and Wi-Fi receiver are operated simultaneously at the higher end of LTE band 40 and the lower end of the ISM band respectively, as shown in Figure 12.5a, the filters provide little or no isolation. In such extreme scenarios, the Wi-Fi receiver must be able to handle high levels of all three self-interference types. On the other hand, the Wi-Fi receiver may only need to deal with leaked LTE transmitter noise for deployment scenarios like that shown in Figure 12.5c.

The required isolation budget allocations for the three scenarios depicted in Figure 12.5 are summarized in Table 12.3. It is assumed that the LTE transmitter is set to the maximum output power (+23 dBm) for all three scenarios. The goal is to ensure that the Wi-Fi receiver desensitization caused by self-interference is limited to 1 dB. Assume that the Wi-Fi receiver employs direct downconversion. It is known that the OOB blocker will create unwanted distortion that ranges from DC to the spectral width of the blocker at the Wi-Fi receiver's mixer output. This distortion is caused by even order distortive processes (e.g., quantified by IIP2) in the direct downconversion Wi-Fi receiver's front end (LNA, mixers) and falls on top of the wanted signal at baseband. Additionally, note that the leaked nonlinear distortion and noise from the LTE transmitter will also interfere with the wanted Wi-Fi signal. Therefore, assuming that the two processes contribute equally, the unwanted interference caused by the OOB blocker and the leaked spurious emission interference from the LTE transmitter must be 16 dB below the Wi-Fi receiver sensitivity level (i.e., −90 dBm for a 20 MHz channel) if the desensitization is to be limited to 1 dB. Assume that the Wi-Fi receiver front-end IIP2 is +14 dBm.[10] Therefore, OOB blocker at the input of the Wi-Fi receiver must be reduced to −46 dBm or lower.

In order to determine the required amount of OOB blocker cancellation, we consider the attenuation provided by the Wi-Fi bandpass filter at the LTE transmitter

TABLE 12.3
LTE Band 40 to ISM Band Isolation Budget Allocations

Scenario	$\xi_{Antenna}$ (dB)	ξ_{LTE} (dB)	ξ_{Wi-Fi} (dB)	Required OOB Blocker Cancellation (dB)	Required Spurious Emission Interference Cancellation (dB)
Figure 12.5a	15	10	4	>50	>64
Figure 12.5b		10	39	>15	>44
Figure 12.5c		50	39	>15	>11

signal's center frequency. For the first scenario depicted in Figure 12.5a, the Wi-Fi bandpass filter provides only 4 dB attenuation (see Table 12.1). Assuming that the isolation between the LTE and Wi-Fi antennas is 15 dB, the leaked OOB blocker level appearing at the input of the Wi-Fi receiver is +4 dBm. In order to limit the second-order (IP2 related) distortion to −106 dBm (16 dB below the sensitivity level of −90 dBm), the OOB blocker must be canceled to a level of −46 dBm if the Wi-Fi receiver IIP2 is assumed to be +14 dBm. Therefore, the required OOB blocker cancellation for the scenario depicted in Figure 12.5a is 50 dB or better. For the second and third scenarios depicted in Figure 12.5b,c, the Wi-Fi bandpass filter provides better than 39 dB attenuation. Therefore, the required OOB blocker cancellation is 15 dB or better.

The leaked spurious emission interference at the input of the Wi-Fi receiver for scenario 1 (Figure 12.5a) is determined by considering the LTE transmitter adjacent channel leakage ratio (ACLR1). The standard's specified minimum ACLR1 requirement is 33 dB.[11] The ANADIGICS ALT6740[12] power amplifier (PA) achieves a typical performance of 40 dB ACLR1 at maximum output power. Note that the LTE bandpass filter provides 10 dB of attenuation in the Wi-Fi channel of interest for this scenario. Therefore, the spurious emission interference from the LTE transmitter appearing at the input of the Wi-Fi receiver is −42 dBm. The required spurious emission interference cancellation is 64 dB. A similar approach may be used to determine the OOB blocker and spurious emission cancellation required for the scenarios depicted in Figure 12.5b,c. Note that the spurious emission created by the LTE transmitter for scenario 2 shown in Figure 12.5b can be determined by considering the alternate channel leakage ratio (ACLR2).[11] The ANADIGICS ALT6740 PA[12] achieves a typical performance of 60 dB ACLR2 at maximum output power. The LTE bandpass filter provides 10 dB of attenuation in the Wi-Fi channel of interest for this scenario. The spurious emission interference from the LTE transmitter appearing at the input of the Wi-Fi receiver is −62 dBm. The required spurious emission interference cancellation is 44 dB. For scenario three, the spurious emission created by the LTE transmitter consists of noise only. The standard's specified maximum LTE band 40 noise level in the ISM band is −30 dBm.[11] The LTE bandpass filter provides 50 dB of attenuation in the Wi-Fi channel of interest for this scenario. The leaked noise in the ISM band is −95 dBn. The required spurious emission interference cancellation is, therefore, 11 dB.

A conceptual block diagram of an RF system that might be deployed to handle coexistence-related self-interference is shown in Figure 12.6. The system is in a state where the LTE transmitter and Wi-Fi receiver are active at the same time. The victim (Wi-Fi) receiver system is shown to employ analog self-interference cancellation (A-SIC) circuitry and adaptive digital self-interference cancellation (AD-SIC) logic. These blocks are highlighted in light gray. Additionally, the context aware RF resource manager (CARF-RM) is shown in dark gray. The CARF-RM is responsible for configuring and dynamically adapting the canceler blocks in a context driven manner. The primary purpose of the A-SIC is to remove the OOB blocker at the input of the main (RX1) and supplementary (RX2) receivers. This prevents the low-noise amplifier, mixer, and analog-to-digital converters employed by RX1 and RX2 from saturating. A copy of both the OOB blocker and spurious emission is made at the output of the aggressor (LTE) PA and fed into the victim (Wi-Fi) receiver interference cancellation blocks as a reference. The A-SIC attempts to recreate a signal

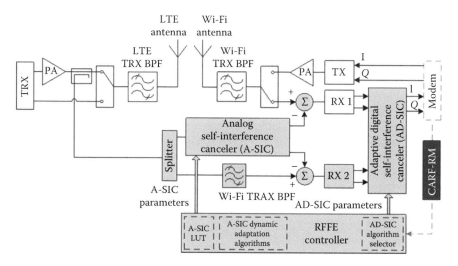

FIGURE 12.6 Radio receiver architecture.

that matches the leaked blocker signal for cancellation. The A-SIC design details are found in references [13,14]. The A-SIC consists of a parallel bank of series connected programmable delay and gain elements. The A-SIC parameters might consists of a set of weights and delays, for example. A lookup table (LUT) in the RFFE controller may contain the initial values of the A-SIC weights and delays for different aggressor and victim carrier frequency combinations over temperature and aggressor transmitter output power levels. Furthermore, the CARF-RM may employ algorithms to dynamically adapt or perturb the A-SIC parameter set to optimize its performance in response to changes in the UE's operating environment.

The AD-SIC removes the residual in-band self-interference (nonlinear distortion and noise) left by the A-SIC or when the A-SIC is inactive. Various adaptive filter techniques (e.g., NLMS) are available for the AD-SIC implementation.[15] The operational states of the A-SIC and the AD-SIC are summarized in Table 12.4. The cancelers are transitioned from one state to another by the CARF-RM.

TABLE 12.4

Interference Canceler States for Various Self-Interference Levels

Self-Interference Level	A-SIC Settings		AD-SIC Settings	
	State	Dynamic Adaptation	State	Performance Level
Level 1	ON	Yes	ON	High
Level 2		No		
Level 3				Low

12.6 CARF-BASED MESSAGING FLOW DIAGRAM FOR IN-DEVICE COEXISTENCE

It is assumed that the LTE RAT is actively running and the Wi-Fi modem to be used for off-loading cellular data traffic. The triggering mechanism could be network based radio resource controller (RRC) messaging, application layer signaling, or another viable way. For example, the user may be active on a LTE network having a voice call and preferred to de-activate cellular data exchange so that an application can initiate data exchange via the Wi-Fi RAT. Also, the RRC messaging can trigger an event to command the Wi-Fi RAT to off-load the portion of cellular data traffic. A high-level control flow is captured in Figure 12.7 that enables the Wi-Fi RAT operations while the LTE RAT is in an active state. The CARF RM based on the provisions allowed either by higher layer messaging or directly accessing predefined register banks that are associated to the RAT state information such as activation state, channel index, bandwidth, and power levels for all RATs. An alternative way to extract the RAT state information is to define a packet inspection entity that may be

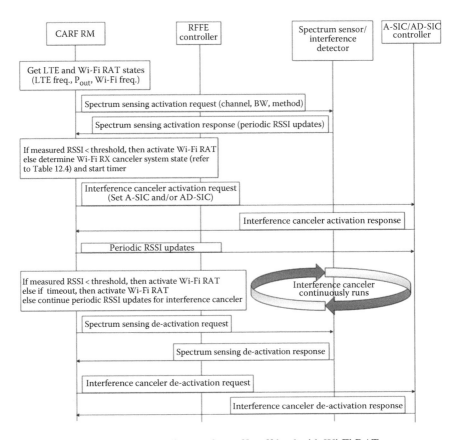

FIGURE 12.7 Messaging flow diagram for traffic off-load with Wi-Fi RAT.

implemented within the modem protocol stack to extract messages that are relevant to the RRC layer targeted for RFFE resources.

In order to enable interference cancellation, the CARF RM utilizes spectrum and interference detector block as depicted in Figure 12.3 to measure the received signal strength indication (RSSI) level where the Wi-Fi RAT will be deployed. Although the Figure 12.7 illustrates the Spectrum and Interference Detector block utilization for the RSSI measurements, the RSSI measurements may, also, be measured directly by using the Wi-Fi protocol stack in test mode if the provisions are implemented. The Wi-Fi modem can be instructed to run under initial state and the RSSI measurements can be periodically generated.

The measured RSSI updates that feed to the cancellation algorithms as shown in Figure 12.7 are continuously generated until either the A-SIC and/or AD-SIC drives the system to an equilibrium that results in less than a target threshold or the reaching of a timeout. The target threshold is chosen such that the Wi-Fi clear channel assessment (CCA) logic may be successful to access the medium. The CCA is one of two carrier sense mechanisms in Wi-Fi systems. It is defined in the IEEE 802.11 standards as part of the physical medium dependent (PMD) and physical layer convergence protocol (PLCP) layer. The CCA utilizes two related functions— carrier sense and energy detect. The CCA carrier sense (CCA-CS) is a mechanism that enables the receiver to detect and decode a Wi-Fi preamble.[16] If the CCA-CS threshold is set to −87 dBm (maximum is −82 dBm for 20 MHz bandwidth), which is +7 dB above −94 dBm that is the noise floor for a receiver utilizing 20 MHz bandwidth with 7 dB noise figure by using Equation 12.1. The Wi-Fi modem by using the PLCP header field extracts the time duration for which the medium will be occupied. The CCA flag is held busy until the end of packet transmission when the Wi-Fi preamble is detected. The CCA energy detect (CCA-ED) enables the receiver to detect non-Wi-Fi energy in the operating channel and back off data transmission. The CCA-ED threshold is typically defined to be 20 dB higher than the minimum PHY Rx sensitivity. The higher CCA-ED threshold setting enables the Wi-Fi systems to be more resilient against non-Wi-Fi signals. However, if the in-band signal energy crosses CCA-ED threshold, the CCA is held busy until the medium energy is below the threshold.

In the case that the Wi-Fi modem activation is reached based on a timeout, it should be assumed that the cancellation algorithms settled to the best possible outcome within the prescribed time and that the measured RSSI value is bigger than the threshold. The failure to settle to below the target RSSI value may be due to other interfering sources rather than the LTE RAT interference measured in the Wi-Fi channel.

Under normal Wi-Fi operations due to the device orientation and proximity as a result of reflective objects being closer to the antennas, the amount of interference experienced by the modem may cause increased medium access delay issues. While the Wi-Fi modem is downloading or uploading data, the CCA failure condition may trigger to run the cancellation algorithms. A way of approaching the issue is to run the interference cancellation algorithms periodically even after seeing convergence to the lowest RSSI value in the desired channel.

12.7   CONCLUSIONS

The component-level enhancements and enabling technologies play an important role allowing the device manufacturers to achieve goals to support ever-increasing requirements motivated by the user demands for ubiquitous, instantaneous over-the-air access to multimedia-rich content.

In order to address increasing data traffic demand and help reduce relevant cellular core network congestion issues, the concept of traffic off-loading has been introduced. The traffic off-loading is achieved by enabling simultaneous operations of multiple RATs. However, the deployment of two separate RATs adjacent in space and frequency may not always be possible. The operation of one RAT with high transmit power levels can desensitize another RAT's receiver in space-limited compact devices when the spectral separation between the two are limited.

In this chapter, we have focused on a system approach, namely, the CARF, to create a framework to enable intelligence in the RF front-end domain. The CARF approach acquires the system-level requirements to optimize the modem's RFFE HW–related resource parameters. Also, we have defined the CARF RM entity to facilitate the RFFE interaction with protocol stack, application layer, user preferences, and other system-level inputs. We have established, as an example, how the cellular data traffic off-loading capability is enabled even when LTE and Wi-Fi RATs are simultaneously operating in adjacent channels. The CARF RM combined the appropriate interference cancellation methods with the feedback provided by the spectrum and interference detector entity in order to reduce the LTE RAT interference over the Wi-Fi channel.

REFERENCES

1. FCC, Third Memorandum Opinion and Order FCC 12-36A1, ET Docket No. 04-186 and ET Docket No. 02-380, 2012.
2. FCC, Further Notice of Proposed Rulemaking FCC 14-49, GN Docket No. 12-354, 2014.
3. FCC, Notice of Inquiry FCC 14-154, 2014.
4. Sankaran, C., Data offloading techniques in 3GPP Rel-10 networks: A tutorial, *IEEE Communications Magazine*, 50(6), 46–53, June 2012.
5. 3GPP TR 36.816 V11.2.0, Study on signalling and procedure for interference avoidance, 2011.
6. Hu, Z., Susitaival, R., Chen, Z., Fu, I.-K., Dayal, P., and Baghel, S., Interference avoidance for in-device coexistence in 3GPP LTE-advanced: Challenges and solutions, *IEEE Communications Magazine*, 50(11), 60–67, November 2012.
7. 3GPP TR 36.814 V9.0.0, Evolved Universal Terrestrial Radio Access (E-UTRA); Further advancements for E-UTRA physical layer aspects, March 2010.
8. 3GPP TSG-RAN WG4 meeting #57 R4-104334, Analysis on LTE and ISM in-device coexistence interference, Motorola, 2010. http://www.3gpp.org/DynaReport/TDocExMtg--R4-57--28052.htm.
9. Bratislava, S., In-device coexistence interference between LTE and ISM bands, 3GPP TSG-RAN WG4 Ad-Hoc Meeting #10-03 R4-102416, Qualcomm Incorporated, 2010. http://www.3gpp.org/DynaReport/TDocExMtg--R4-56--28051.htm.
10. Emira, A.A. et al., A dual-mode 802.11b/Bluetooth receiver in 0.25 um BiCMOS, *ISSCC*, San Francisco, CA, Vol. 1, pp. 270–527, 2004. Doi: 10.1109/ISSCC.2004.1332698.

11. ETSI TS 136 101 V10.3.0, LTE; Evolved Universal Terrestrial Radio Access (E-UTRA); User Equipment (UE) radio transmission and reception, 3GPP TS 36.101 version 10.3.0 Release 10.

12. Data Sheet—ANADIGICS ALT6740 power amplifier for LTE devices operating in LTE band 40.

13. Bharadia, D., Mcmilin, E., and Katti, S., Full duplex radios, *SIGCOMM, ACM SIGCOMM Computer Communication Review*, New York, Vol. 43(4), pp. 375–386, 2013.

14. Zhou, J. et al., Low-noise active cancellation of transmitter leakage and transmitter noise in broadband wireless receivers for FDD/co-existence, *IEEE Journal of Solid State Circuits*, 49(12), 3046–3062, December 2014.

15. Haykin, S., *Adaptive Filter Theory*, Prentice Hall, Upper Saddle River, NJ, 1996.

16. Prehia, E. and Stacey, R., *Next Generation Wireless LANs*, Cambridge University Press, Cambridge, UK.

13 Network Operator Perspectives

Yuang Lou

CONTENTS

Long Term Evolution (LTE) mobile services started its deployment at the turn of 2010 and 2011. By the end of 2013, there were 268 LTE networks launched in 100 countries with a total of 157.7 million LTE users[1] worldwide. Among these, 112 million LTE users have been added in 2013 and 79 million of them were activated from the North America markets. This growth illustrates a strong growing momentum for the LTE radio platform with an all-IP core network infrastructure to improve and expand the mobile business opportunities globally.

Presently, mobile network operators are diligently exploring opportunities for LTE deployments with higher business potentials and new services. The mobile communication community wants to understand the foreseeable challenges in LTE advancement and required regulatory innovation and technical support to enhance LTE service deployment. Based on business expectations and market developments, LTE RAN equipment and device vendors want to align their R&D activities and technical investments not only to resolve the performance challenges and improve the user quality of experience (QoE) but also to support both architecture evolutions in future LTE RANs and mobile devices.

13.1   LTE DEPLOYMENTS AND CHALLENGES

Figure 13.1 demonstrates the worldwide mobile network coverage where LTE is one of the booming 4G service options. The World Time Zone website [2] further lists the detailed distribution of LTE capabilities and coverage.

According to Cisco's annual report,[3] mobile data traffic has been continuously growing in the past few years and a similar trend will remain in the years to come. In North America, the mobile data traffic is predicted to grow eight-fold by 2018. During the same time period, global data traffic growth in mobile communications will increase nearly 11-fold. From a long-term view, expected growth in mobile traffic data could eventually reach 1000 times[4,5] what we have now. This growth demonstrates an excellent opportunity for ongoing LTE development but also raises challenges and concerns in many aspects.

To support higher data volume and faster mobile traffic, more RF spectrum resources are required. Based on current market needs and spectrum allocations, improvements in spectrum utilization and spectral efficiency become urgent for both of the downlink and uplink mobile operations. This presents a serious challenge to existing RF spectrum allocations. And both of these improvements will eventually translate into technical challenges that will further shape the evolution and innovation in mobile communications. For example, even if there are many viable spectrum opportunities such as the worldwide digital dividend review[6] (DDR) developments to convert RF bands from analog TV broadcasting to mobile Internet accesses in 600 MHz or the technical innovation spectrum in 3.5 GHz for small cell–HetNet deployment, the regulatory process, auction activities around RF spectrum, and legacy service migration following new regulatory rules will take time, which is highly pressured by growing deployments of LTE services. Beyond these licensed spectrum options, Authorized/ Licensed Shared Access (ASA/LSA[7,8]) has been developed. Related RF spectrum regulatory support has been moving forward and becoming more open. Under ASA/LSA operation, shared spectrum access can be permitted in the markets where additional primarily licensed spectrum is not yet available.

Wireless data access via laptop computer networking on unlicensed RF spectrum have been developed and have become popular forming a complementary coverage to the consumer data access from the wired cable network—either physically built on top of copper-based Ethernet or on top of fiber-based optical infrastructure. Unlicensed RF data access extends the device's network connection from a fixed access point to a tethering, nomadic access area and growing toward a mobile access connection. Supporting a mobile data network access has become the most important and unique capability embedded in the wireless communications.

Unlicensed wireless connections have increased dramatically with data accesses from desktops, laptops to tablets, and even mobile phones. It forms an effective venue to off-load mobile traffic flow from licensed macrocellular network to the unlicensed Wi-Fi access. As the referenced study[3] shows, there was at least 45% of total mobile data traffic off-loaded from cellular network to Wi-Fi or femtocell networks in 2013 globally, at an averaged off-loading level of 1.2 exabytes of data volume per month. Without this off-loading support, mobile data traffic would have grown 98% rather than 81% annually in the year, which would greatly challenge the existing capacity

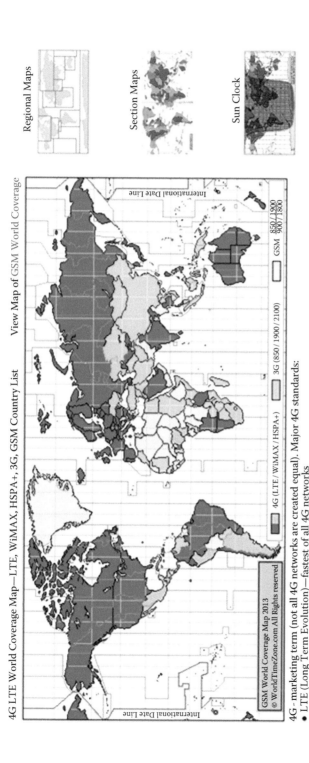

FIGURE 13.1 Worldwide coverage of mobile communications services. (From http://www.worldtimezone.com/4g.html, updated February 12, 2014.)

of the mobile networks and heavily impact the QoE performance of consumer data access. In order to meet the challenge of 1000× growth in the user data traffic, small cell–HetNet complemented with unlicensed Wi-Fi mobile access is an important option in the network architecture.

Data off-loading is the only one part of developments in future mobile data access. With data off-loading, a mobile user session may not be continued during the access change. The device has to stop the existing session and start a new one. As long as the switching latency from the data off-loading is short enough, users may not even be able to sense a session interruption. On the other hand, the off-loading scheme will challenge the QoE of mobile streaming services. The annual report shows that 50% of mobile traffic is from mobile video access.[3] Mobile user behaviors have been continuously shifting to IP video access. Instead of reading the news in text format from a mobile device, news streaming through mobile audio/video delivery attracts users' attentions. For improved QoE, both audio and video streaming services require seamless session continuity instead of switched data off-loading between the licensed and the unlicensed network accesses. QoE enhancements, enriched mobile services, and seamless mobile video streaming delivery drive the need for innovations in spectrum regulation and technical advancements in user radio access.

In the past few years, service providers have made progress with new LTE service offerings on top of the expanded coverage of LTE network access. For example, the LTE SON deployment and mobile device access capabilities to Wi-Fi networking[9] were both initiated in the past few years; efforts to deploy small cell networks and Wi-Fi off-loading started in 2013; at the beginning of 2014, mobile application of TV everywhere[10] enables live TV contents to be delivered to and displayed on user's smartphones; services of carrier aggregation (CA)[11,12] and VoLTE[13] have been launched in markets one after another in the first half of 2014. Looking into the future, it is not too hard for us to envision that there will be more mobile business opportunities and deployed service features over LTE and LTE-Advanced platforms to support medical applications carried over LTE, LTE-enabled broadband vehicle-to-vehicle communications, dynamically reconfigurable CA bands that logically bind more RF bands into one wider bandwidth operation, and so on. The LTE radio platform becomes a flexible but core foundation to future mobile broadband data communications. The IP multimedia subsystem (IMS)[14] architecture bridges all of the IP multimedia services. More software-defined capabilities will be deployed in future LTE networks. Technically, software-defined network (SDN) characterizes the network evolution for virtual network functions and dynamic network management.

13.2 MOBILE DEVICE SUPPORT TO ADVANCED LTE RADIO PLATFORM

The service platform of mobile Internet communications has evolved with the transitions from analog to digital, from circuit-switched to packet-switched, and from narrowband to wideband mobile radio operations. A wideband radio platform enhances higher data transmission rates with multiuser access simultaneously, but it faces several open challenges and requires an agile and flexible mobile radio architecture.

One of the challenges to the advanced radio platform is how wide an RF bandwidth that an LTE mobile architecture can support and where/how we can have a wide and even wider RF bandwidth to support the operation. Frequency-selective fading challenges an efficient and reliable network operation due to the existence of an inherent irreducible error floor from the setup of wideband radio channels.

Compared to 2G[15] and 3G[16] mobile systems, higher level QAM modulation in 4G LTE[17] across multiple resource blocks (RBs) raises the system peak-to-average-power ratio (PAPR) challenging the battery power demand and supply in mobile devices. OFDMA operation in the LTE downlink is not suitable to that of the LTE device uplink transmission when an optimized and efficient device operation is considered and configured.

Facing the aforementioned challenges, OFDM[18]-based LTE becomes an ideal wideband radio platform. Its higher rate delivery can be achieved with simultaneous transmission over many orthogonal RF subcarriers.[19] Its OFDM architecture breaks the wideband frequency-selective fading into many narrow but non-frequency-selective and flat spectrum characterized subchannels. In this way, higher system performance can be achieved through aggregated narrowband operations. A wider RF bandwidth can be configured in a flexible and scalable fashion from 1.4 MHz up to 20 MHz block. In the market places, a continuous 20 MHz RF block pair is rarely owned by any FDD mobile network operators due to competitive spectrum auctions. It forms another challenge to offer 20 MHz LTE bandwidth through the network operation. The LTE-Advanced platform can support both logically combined adjacent and even physically separated intraband and/or interband RF blocks into a wider-channel bandwidth not only up to 20 MHz but also up to 100 MHz via CA. CA resolution becomes a key advancement to expand wideband LTE operation while more software-defined features are expected to run on top of it. For example, CA capability can be defined by a specific band pair combination in 3GPP Band 17 and band 4. Advanced CA is able to reconfigure more combined band pairs dynamically according to the consumer demands and also depending on the available spectrum resources in the field operation.

To properly cover the differences and complications between LTE base station and device operations, LTE adopts the SC-FDMA[20]-based mobile platform in its uplink operation, which runs a simplified device processing to limit the PAPR.[21] It reduces the device battery power consumption as compared to that of downlink OFDMA. Beyond the support of wideband operation, the LTE radio design is specifically focused on the enhanced implementation of antenna array processing in terms of raising the RF spectral efficiency. MIMO[22] transmitter and receiver over an antenna array is specified in both downlink and uplink of LTE operations. Varied MIMO device operations, adaptive modulation and coding schemes (MCS), and CA make the 4G LTE platform fit in wideband communications much more efficient. In essence, 4G LTE is characterized by an all-IP-based system with a flat network architecture, a dynamic RF link adaptation guided by the instant MCS selection, and an IMS-based core to bridge the legacy mobile operations across multimedia services with MIMO over both OFDMA downlink and SC-FDMA uplink radio platforms.

Looking at the current LTE deployments, we realize that, among many already achieved performances, other advanced features need to be developed. For example,

Item	Subcategory	IMT-Adv	LTE
Spectral efficiency (b/s/Hz)	Downlink	15 (4×4 MIMO)	16.3 (4×4 MIMO)
	Uplink	6.75 (2×4 MIMO)	4.32 (64QAM)
Data rate (Mbps)	Downlink	1000	300
	Uplink	100	75

FIGURE 13.2 LTE versus IMT-Advanced. (From http://www.home.agilent.com/upload/cmc_upload/All/23Jan14WebcastSlides.pdf.)

defined in the 3GPP, the goal of LTE RAN radios is to support 8×8 MIMO and the LTE device is to support 4×4 MIMO operations, not to mention the performance differences between LTE and IMT-Advanced shown in Figure 13.2 with LTE-Advanced capable of meeting and surpassing the IMT-Advanced specifications. In current field operations, deployed LTE RAN radios can only support 2×2 MIMO and the LTE mobile device radios can only support 1×2 MIMO. It demonstrates a big performance gap between the 3GPP goal and the reality from field operations while simultaneous radio transmissions in the same mobile device are still challenged by both the battery power consumption and inter-radio interference. The goal to achieve higher LTE spectral efficiency via a higher-order MIMO is still on the development agenda.

The latest smartphone design enables multi-active radios to support LTE and Wi-Fi radio transmissions simultaneously. Further development shows that the 2×2 MIMO device chipset for IEEE 802.11ac has been commercially released[23] in early 2014. The 2×2 MIMO and 4×4 MIMO LTE antenna arrays were already prototyped in mobile devices.[24] Their performances have even been evaluated and compared through market validations. All of these achievements make the device operation in higher-order MIMO characterized by the simultaneous LTE radio transmission/receiving more attractive in commercial deployments. OFDM-based multiuser-MIMO[25] (MU-MIMO) opens another venue to raise the spectral efficiency through the space-division multiple access (SDMA) in the network access. It also requires mobile device support. Even though the initial MU-MIMO is only applied to the access of IEEE 802.11ac, its capability should be extended to improve LTE spectral efficiency as well.

Beyond raising the spectral efficiency to LTE cell coverage, cell edge performance attracts the attention of network operators. This interest is strengthened by small cell–HetNet deployments. Small cell coverage and macrocell coverage are overlapped (see Figure 13.8), which raises the total network data capacity via a preferred data off-loading to the small cell access. Enhanced small cell operation at the cell edge off-loads the traffic volume from the macrocell and thus opens more macrocell resources to handle higher mobility accesses. Many more LTE-Advanced features[26,27] are going to be developed, such as SON, FeICIC, and coordinated multipoint (CoMP), which further require support from advanced mobile devices.

Future radio development will not be purely hardware oriented. Software-defined capabilities characterize LTE evolution. Particularly with mobile devices, limited device volume pushes its development toward the improved operational efficiency with intelligent control in its radio access. Advanced LTE mobile device performances are fully reflected in the following areas: enhanced multimode multiband (MMMB) access capability with device volume efficiency, battery power efficiency, and RF spectral efficiency.

13.3   BUSINESS OPPORTUNITY WITH LTE ADVANCEMENTS

Initial LTE deployment has been focused on mobile data access only. Simultaneous voice capability to an LTE user is either provided with an additional CDMA radio or employing circuit-switched fallback (CSFB) to the 2G or 3G voice services. Evolved with network advancements, mobile devices have been enabled access to not only international data roaming[28] with varied RF band options fitting to local market demands, but also to VoLTE capabilities. VoLTE over the cellular platform is a preferred voice option for the best consumer QoS. It is not only an IP-based voice service but also IMS-based, which serves as the common service core to support all of the legacy multimedia operations.

As mentioned earlier and shown in Figure 13.3, mobile video demand has been surging as reflected in the growth of mobile data traffic. Unicast mobile video streaming delivery is a service option to meet certain users' demands. In addition to unicast services, there is an economic and capacity opportunity to mobile multicasting and broadcasting through the eMBMS video/TV delivery. Challenges here are not only the deliverable quality of mobile video streaming but also the market support for the delivery of mobile video streaming. In a conventional broadcasting system, video content is timely scheduled to broadcast no matter whether there are few viewers or even

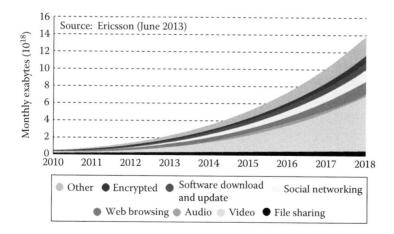

FIGURE 13.3   Sixty percent of annual growth of mobile video consumption. (From http://www.cedmagazine.com/news/2013/06/mobile-video-consumption-growing-at-60-percent/year.)

no viewers in the service area at all. In the past MediaFLO operations, the broadcast spectrum, network infrastructure, and user mobile devices were all standalone and not shared across any other mobile services. Merging mobile video delivery with LTE services and sharing the LTE spectrum/devices can overcome economic barriers in offering mobile TV broadcasting. Dynamic reconfiguration via capable software control should make eMBMS broadcasting more efficient and flexible together with LTE radio resource management to meet the consumer demands. Shared spectrum, networking, and mobile devices between eMBMS and LTE[29,30] infrastructure with switchable unicast, multicast, and broadcast service model will attract the future service development while the supplemental downlink (SDL) spectrum has the potential to support both eMBMS and CA operations. Mobile video delivery and LTE operation should not be limited by the OFDM architecture but be open to a dynamic reconfigurable operation on an OFDMA basis following market demands. This new model of mobile video streaming and TV broadcasting as part of the LTE operation demonstrates the opportunities and also challenges to the network configuration and mobile device support.

The LTE network will continuously support the human connections via voice and data communications but could also be extended to web-based medical service coverage. It is evolving to support wireless machine-to-machine (M2M)[31] communications for network monitoring and remote control under the Internet of Things (IoT). It will also be enhanced to support mobile vehicle-to-vehicle (V2V) communications for connected cars via LTE[32] in order to strengthen the in-vehicle security, hands-free call, turn-by-turn navigation, and remote diagnostic services. Within the IoT access, a large percent of communication traffic will be between fixed equipment, between mobile devices and fixed equipment, and also between mobile-to-mobile devices. To support all of these low power but highly dense and instantly connected devices under the LTE management is another challenge and opportunity to both the innovative network configuration and mobile device architecture.

The success of Internet and cellular communication has been merged and raised onto the track of cloud computing (CC) for data applications, video streaming delivery, and IoT services over strongly supported mobile platforms. Three major CC developments, such as the cloud-based data storage, cloud-based online processing, and cloud-based network services, will be accelerated under the user-initiated SDN operation where the network equipment can be flexibly managed and reconfigured by the virtualized functions and enriched service features with more intelligence but less redundant hardware equipment. Building up a "User-Defined Network Cloud[33] (UDNC)" becomes the goal of the mobile network operator who puts user satisfaction at the top priority in network operations. Its MMMB device access and self-reconfiguration performance directly impacts the user QoE and becomes one of the keys to qualify the UDNC operation.

13.4   CHALLENGES TO ADVANCED LTE MOBILE BROADBAND DEVICES

The operation of mobile RAN equipment focuses on its designated RF band and radio protocol but shares implementations across the hardware platform in only very limited ways. For example, 2G, 3G, and 4G RAN configurations are independent of each other from the antenna array of the air interface down to the RF transceiver units. Separated RAN operations handle their own traffic flow, IP signaling[34] to the backhaul service, mobile radio access, and user data delivery.

Radio architecture in mobile devices is totally different from that of the RAN. Service options coexist in the same mobile device to support the unique end point performance of user QoE delivery. Capable mobile devices support MMMB operation. Supporting a higher number of MMMB access is one of the merits in the device capability such that a customer-desired device should be able to access multiple services from 4G LTE data, simultaneous 2G/3G voice call, GPS tracking, and the delivery of mobile video streaming, to just name a few. Widely expected service capability forms specific challenges to the LTE device antenna array, front-end module (FEM), RF transceiver, mobile operations for interference control, and effective battery life management.

In the United States, the RF spectrum for 2G mobile operation is defined from 850 MHz to 1.9 GHz. The 3G RF platform covers the same frequency range of 2G but extends its high frequency end up to 2.1 GHz. The spectrum coverage of LTE is much wider—from low 700 MHz to 2.6 GHz, which will be further stretched down to 600 MHz and also extended up to 3.5 GHz. Wider RF coverage becomes one of the characteristics of 4G operations. This wide spectrum coverage challenges the configuration of device antenna to 4G LTE, which requires a multielement antenna array to support the MIMO operation. Beyond a MIMO antenna array, more than one antenna is also required to cover the RF range for different band access. On average, there are four to five antennas in an LTE smartphone to cover bands, modes, and many other service options, such as GPS, Wi-Fi, FM radio, NFC, and 2G/HSPA/ LTE accesses. Due to limited device volume, the same cellular mobile antenna is usually shared among 2G/3G and LTE bands. For example, 3GPP band 4 in 2.1 GHz and band 17 in 700 MHz can share the same piece of antenna. Inserting an additional antenna normally gains in its Tx/Rx efficiency and RF isolation but is challenged by the device volume. Envelope correlation among array elements is sensitive to the distance separation of antenna elements. Sharing the same antenna across bands improves the device spacing efficiency but challenges the achievable RF performance. Based on all the feasibilities and compromises, tunable antenna becomes a key contributor to the device antenna design in volume, element spacing, and high-RF performance. Its contribution is more important to lower frequency operation where physics require a larger antenna size proportional to the wavelength in terms of its better RF performance.

13.5   TUNABLE DEVICES WITH MIMO ANTENNA ARRAY

Fixed antenna designs at the hardware level cannot optimize their wideband operation to cover all RF bands. In other words, antenna hardware performance in some bands is better than that in others. Dynamically adjusting antenna resonant frequency or matching antenna impedance is the way to improve antenna efficiency but requires intelligent software control. Noticeable challenges are particularly evident in lower frequency operation such that, in the 700 MHz band, the device antenna gain is generally low (around −7 dBi) from the compromise of a wideband but small-sized antenna design. The gain gap between low- and high-band antenna performance is usually greater than 6 dB on the same piece of broadband antenna. The receiving gains of MIMO array antenna elements should also be narrowed down. If the Rx gain difference among the elements is >3 dB, it impacts the performance of MIMO receiving process.

Improved performance over a wideband antenna is achievable through software reconfiguration either via open- or closed-loop tuning. Open-loop tuned antennas have already been deployed in several mobile devices. Varying input feeds could be used to tune the antenna resonant frequency. Adjusting digital capacitance can match the antenna impedance and thus raise the antenna efficiency. Up to this point, open-loop tuned antenna demonstrates improved total radiated power/total isotropic sensitivity (TRP/TIS) performance on a single antenna element.

Closed-loop tuned antennas are more complicated than that of the open-loop ones. Closed-loop tuned antennas have to detect and track antenna detuning and then match with the best impedance to compensate for the detuning. The process of detection and best matching is adaptive and continuous. In order to reach large-scale commercial deployment, solutions of closed-loop tuned antenna need further technical improvement and market validation.

The basic assumption to the current open- and closed-loop tuned antennas is that there is only one RF band or mode actively transmitting and receiving among the many available but inactive. To improve LTE MIMO performance, network operators are interested to see enhanced tuning in the antenna array instead of TRP/TIS improvements over a single antenna element. To be more specific, a valid MIMO antenna tuning will not only focus on the individual TRP/TIS across all antenna elements, but also includes narrowed Rx gain difference and reduced envelop correlation across all array elements together with the best TRP/TIS at the same time and even across the CA bands. Less differences in receiving antenna gains across all array elements improves the MIMO processing. A joint MIMO antenna tuning, including reduced envelop correlation across array elements, increased RF isolation and minimized RF interference among the elements together with the best achievable TRP/TIS, boosts the overall performance on all MIMO antenna array elements. Achieving joint performance of the antenna array is the goal of tunable MIMO device antennas.

Current implementation of 1×2 MIMO antenna arrays in mobile LTE devices needs qualified tuning support to cover both primary and secondary receiving antenna elements. A higher-order MIMO chipset such as the 2×2 MIMO[23] has been released in the early 2014 to IEEE 802.11ac. Device array antenna architectures to support 4×4 MIMO have gone through an initial market evaluation.[24] Both of the aforementioned higher MIMO cases are implemented in the GHz range. Without a matured device antenna tuning resolution, it is hard to deploy 2×2 MIMO into the lower frequency range, that is, to the device operation below 900 MHz. Table 13.1 demonstrates a business forecast in 2013 on tunable device antennas as an open but growing opportunity.

Tunable antennas are active RF components attached to the device FEM. Beyond the potential benefits from antenna tuning, the tuner itself could possibly generate RF harmonics and intermodulation products that would impact spurious emission,

TABLE 13.1
Business Forecast on Tunable Antennas

	2011	2012	2013	2014	2015	2016
4G (LTE, LTE adv) units (M)	17	81	203	324	476	635
Primary antenna tuner ($)	—	0.60	0.90	0.80	0.73	0.56
Diversity antenna tuner ($)	—	0.00	0.50	0.50	0.45	0.42
LTE-Adv incremental ($)	—	0.00	0.00	0.00	0.60	0.70
Diversity tuner attach rate	—	—	0.25	0.5	0.65	0.75
Antenna tuning business ($M)	—	48.6	208.1	340.2	772.4	1061.1

Source: Oppenheimer & Co. Inc., New York, March 2013.

increase insertion loss (IL), and narrow the antenna's bandwidth. These side effects can curb and challenge the benefits gained from the antenna tuning for flexible RF band options and combinations in the mobile devices. Providing a full-scale evaluation on the tunable MIMO antenna array is important to network operators for their device performance support and also important for spectrum regulation. This is a desired technical outcome but may still need more engineering efforts to accomplish.

13.6   RECONFIGURABLE DEVICE DUPLEXER/FILTER

Figure 13.4 shows an example of a common FEM architecture in mobile devices with a bank of device duplexers bridging the RF chains for each individual band of operation that shares a piece of wideband antenna via the fan-out of an antenna switch module (ASM).

Without the capability to reconfigure the duplexer into different operating bands, raising the number of MMMB device access will result in larger duplexer banks. This challenges the device volume and also the performance of the device FEM. Each duplexer is in essence a pair of RF filters permitting the in-band signal flow through but rejecting the out-of-band interference in order to protect its own two-way communications. A qualified duplexer reconfiguration has to meet both of the central frequency shift and OOBE requirements specific in each band. It is a more complicated process compared to the frequency tuning of a MIMO antenna array. Current technology is available to make a reconfigured notch filter across the frequency band but not matured enough to support a cross-band reconfigurable duplexer in general.

Some early results in reconfigurable duplexer have been reported[35] recently with limited working range. To provide a fully reconfigurable and commercialized duplexer requires further R&D efforts.

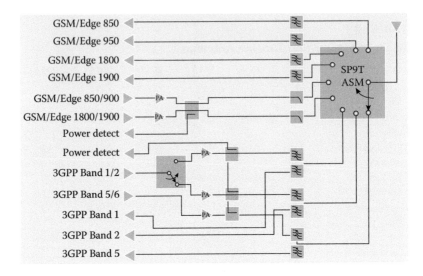

FIGURE 13.4   Duplexer bank connected to the ASM.

13.7   EVOLVED AND INNOVATED DEVICE POWER AMPLIFIER

Power amplifiers (PAs) are key elements in MMMB devices. PAs consume ~50% of the total device power and hold various expectations and potential for future improvements. Advanced PA developments have continuously been evolved and more innovations are coming. Similar to that of the duplexer, the PA performance is also band specific. More MMMB accesses require more PAs in current device FEM. To share a common PA-biased circuitry and input/output ports for a reduced bill-of-materials (BOM), hybrid PA modules (PAMs) have been developed. GaAs PAs have been a matured technical option for a short turnaround time to the market from design to the manufacturing. GaAs PAs have excellent RF performance and reliable operations but are not easy to support the evolution to an integrated FEM. CMOS-based multimode PA (MMPA) has emerged to lead the device FEM integration. At this moment, even though the performance of CMOS MMPA does not exceed that of GaAs PAM in high-power operations, integration of the device FEM with improved PA performance has been a strong incentive to motivate the development of CMOS MMPA.

Figure 13.5 illustrates a view of the CMOS FEM integration within mobile devices, where CMOS MMPA only demonstrates one dimension of the PAM evolution from advanced PA design and silicon processing. Another aspect of device PAM evolution targets for its reconfigurable capabilities based on an intelligent platform under both software and hardware control. It supports the dream of sharing the same device PAM to enable MMMB operations. Early reports on a reconfigurable PAM[36,37] were encouraging and attractive. Network operators are interested to see more advancement with further spacing efficiency and power efficiency improvements leading to an integrated and reconfigurable device PAM. Interest continues along with its expanded spectrum coverage.

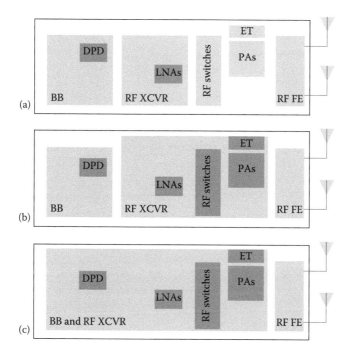

FIGURE 13.5 Evolved SoC options for CMOS-integrated device FEM, agile device RFIC, and SDR: (a) current level of silicon integration for a typical MMMB handset modem, (b) possible future level of silicon integration of MMMB CMOS SoC for handset modem, and (c) "ideal" level of MMMB CMOS SoC for handset modem.

The concept of tunable and reconfigurable RF implementation can be applied to device RFICs. In the current smartphone, RFICs are multiport units responsible for the RF transmission and reception to support device MMMB operation. These ports are generally designed and configured to serve the low (<1 GHz), the mid (1–2.3 GHz), and the high (>2.3 GHz) band options. Multiport RFICs demonstrate agile capabilities to support a set of band selections and configurations through the OEM device design. One of the challenges to RFIC is that required total MMMB access is continuously increasing, say from a 7-band support to the 12-band capable. From an implementation point of view, the total number of RFIC ports cannot be expanded without a limitation. In essence, the cost of RFIC packaging limits the growth of its total ports. Software-defined capability is expected to support port reconfiguration. If one RFIC port can be software reconfigured into different modes on the fly during device operation without increasing the total number of RFIC ports, the same RFIC could double the access capacity to the modes and bands. Similar to the reconfigurable efforts in mobile device duplexers and PAMs, reconfigurable RFIC ports are also band and mode specific. Based on the current development, we have seen that the tunable and reconfigurable evolution on individual RF components in the device FEM, including the agile RFIC, eventually leads to a software-defined radio (SDR)-like mobile device architecture that can coordinate all the reconfigurable capabilities under one central control.

SDR is not a pure software solution. It requires mutual support from both hardware and software. The hardware forms an available FEM platform to support the software reconfiguration for a jointly achievable Q and OOBE in duplexer and PAM, and is also equipped with temperature-compensated characteristics to meet the bandwidth requirements particularly in the UHF operations in 600–700 MHz or even below. More specifically, SDR can only be implemented in the digital domain and relies on ADC/DAC signal conversion across domains. Current ADC power consumption limits its performance in its direct sampling of analog RF signals. At this point, a full-scale SDR expectation serves as a goal of evolution directing the progressive developments of tunable and reconfigurable agile radios in the mobile devices.

13.8   POWER PERFORMANCE IN MOBILE DEVICES

Power efficiency has been a long-term and continuous challenge along with the progressive achievements in device spectral efficiency and spacing efficiency. To limit increased PAPR[21] resulting from higher QAM modulations and to simplify the mobile device operation, SC-FDMA has been adopted in LTE device operation that reduces the device power consumption. Beyond the power efficiency in mobile transmission, battery quality defined in W*h/L, power efficient device processing, and effective power management are three power pillars to support the device operation.

In a typical layout of a mobile device, about 60% of the total device area is reserved for the battery supply (see Figure 13.6). Increased battery volume followed by the size growth of device display contributes to the increased battery capacity from 1420 mAh in the iPhone 4 with a 3.5 in. display up to 3200 mAh in LG G Pro2 with a 5.9 in. display. Even though the battery volumetric energy density has been

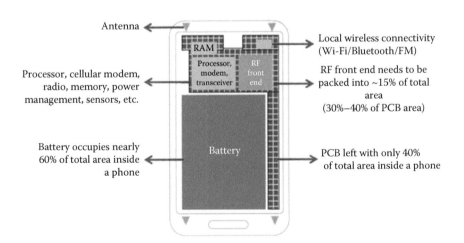

FIGURE 13.6   Typical layout of current smartphone design. (From Carson, P. and Brown, S., *Microwave J.*, 56(6), 24, 2013.)

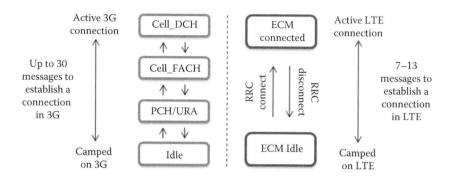

FIGURE 13.7 Compared state models in 3G and LTE radio architecture mobile devices with multi-core processing. (From Lou, Y., Evolved Mobile Broadband Services and Devices, Presentation to *IWPC Workshop on Optimizing Advance Smartphone and Tablets Architecture*, March 19, 2014, http://www.iwpc.org/ResearchLibrary.aspx?ArchiveID=210& Display=doc#.)

more than quadrupled from 150 to 730 W*h/L[38] through innovated technologies, the power gap between the mobile device demands and improved battery supply remains open and has even been widened. Parallel efforts to the device battery improvements in W*h/L and in raised battery size are also shown in the developments from improved device power management (PM), effective radio protocol control, and drastically reduced device PA heat dissipation.

In order to shorten the response latency and improve the QoE in device access, mobile devices have to maintain a proper power level attached to each radio state during the device idle periods, which directly impacts device power consumption. The power level of radio states and the switching timer are controlled by the network scheduler on the RAN side. To understand the difference in radio states between 3G and LTE systems, Figure 13.7 shows a simplified LTE radio state design that is much flatter with less layers to achieve efficiencies in radio resource control (RRC). Connected state discontinuous reception (cDRX) is one of the LTE RRC features. In the cDRX mode, an LTE mobile radio periodically suspends its RF transmission and receiving but continuously holds the allocated RF resources and network connection. Beyond gaining a shortened latency, an LTE mobile device radio is fast to respond with less signaling flow and lowered battery power consumption. A study shows[39] that the cDRX is critical to support VoLTE applications for an effective device battery power management.

Studies on mobile device power consumption show[40] that about 50% of device power is spent on the PAM operation. Most of the PAM power is wasted through heat dissipation, which results in a limited PAM power-added efficiency (PAE). A joint process of averaged power tracking (APT) and envelope tracking (ET) has been developed[41] to reduce the heat dissipation and raise the PAE through the PAM operation. An analog reference interface to ET (eTrak) has been specified in the MIPI standard,[42] which not only contributes to effective device power management but also ensures extremely low-noise performance to reduce the RF interference to the cellular radio receiver.

Prior to 2011, the design of mobile devices was dominated by single-core CPU architecture. Raising both the core processor rate in GHz and bias voltage to the CPU core processor could increase the device processing power but challenge the battery life since power consumption in the core processor is proportionate to its GHz rate and the square of its bias voltage.

Raising the processing power without reducing the device battery life is one of the expected goals in mobile data processing in order to enable multitasking capability and also to enhance the audio- and video-related user applications such as the support of mobile gaming. Dual-core CPU architecture was first introduced into mobile devices in 2011. Power consumption and coordination of multi-core processing were the major concerns. Later studies show[43,44] that, beyond gaining the processing power via the multi-core CPU architecture, dual-core CPUs improve the device power budget by 40% relative to that of the single-core device architecture. This achievement is gained by not raising the CPU processor rate in GHz and not increasing the bias voltage to the core, but distributing the processing workload evenly through a coordinated parallel core processing.

Progressive developments are followed by more cores added to the device architecture. Quad-core CPU devices were initiated in 2012, and 8-core CPU with 64-bit processing has appeared within high-end mobile devices in later 2014 and at the beginning of 2015 such as the incoming Samsung Galaxy A7.[45] The multi-core CPU approach not only enables parallel processing with a relatively low rate processor support but also limits the bias voltage to the CPU processor. Along with the advancement of multi-core CPU device architecture, additional cores are allocated for the multi-core power management and processing coordination.

The development effort of multi-core processing has been supported by the mobile's OS. Current mobile OS is open to support enhanced data processing, efficient core power management, and multitasking-enabled devices up to the 64-core CPU architecture.[46] It forms another branch to the evolution of future mobile devices.

13.9 IMPROVED ACCESS CAPABILITY VIA ENHANCED NETWORK ARCHITECTURE

As to the foreseeable future, the growth of 1000× mobile data traffic[5] has been expected from combined services of broadband communications and the IoT over wireless operations. Based on limited spectrum resources, small cell deployments to reuse the spectrum and to raise the network data capacity under macrocell coverage form the heterogeneous network (HetNet). The HetNet encourages the off-load of user access traffic from the macrocell to the small cell but creates challenges to reduce the RF interference on the signaling flow and data channels to the unintended mobile receivers. To support the real-time mobile applications, such as mobile video delivery and TV broadcasting, additional challenges to the HetNet will include the maintenance of smooth session continuity between small cell and macrocell accesses. Beyond the network configuration, the performance of mobile devices directly impacts the consumers' QoE. The capability of mobile devices to cancel

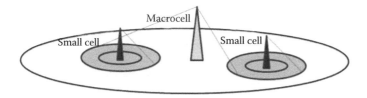

FIGURE 13.8   HetNets view: Small cell deployed within macrocell coverage.

RF interference and coordinate the small cell and macrocell accesses is a key to the HetNet success (Figure 13.8).

Within the macrocell coverage, many small cells could be configured and deployed sharing the same RF spectrum. It requires capable mobile devices to identify and reliably receive its own session data, but reject the RF interferences from the neighboring user accesses and maintain a high quality of receiver performance. This capability requires coordinated efforts between the mobile devices and the HetNet access network in both time domain and frequency domain operations.

For example, intercell interference coordination (ICIC)[47] is defined in the frequency domain and mainly focuses on access interference control over the mobile data channels. Enhanced ICIC (eICIC)[48] and further enhanced ICIC (FeICIC)[49] have been defined in both time and frequency domains where the cross-carrier scheduling and almost blank subframes (ABS) are focused on interference control over the mobile signaling flows. The SON implementation of CoMP[50] operation improves the network capacity, flexibility, and reliability.

LTE license-assisted access (LAA)[51] opens another venue to raise user access capabilities from the evolution of network architecture. All of the aforementioned efforts encourage mobile devices to off-load the traffic from the macrocell to the small cell and unlicensed spectrum access once they are available. This can improve data capacity, though the use of unlicensed radio bands comes with its own challenges and is not a substitute for additional licensed RF spectrum.

13.10   INNOVATIVE SPECTRUM POLICY LEADS TO FURTHER MOBILE FEM EVOLUTION

Improving network capacity to meet the growth of mobile access traffic is a multi-dimensional effort from the HetNet configuration, enhanced spectral efficiency via MIMO, and also at the cell edge through CoMP. It requires innovative spectrum allocation policies in regulations. ASA/LSA complements the small cell off-loading, which enables market-dependent use of military bands, forming secondary access[52] configuration.

Separated device antennas and RF frequency bands between current LTE and Wi-Fi configurations provide enough RF isolation to permit simultaneous radio transmissions in the same mobile device. Inserting more unlicensed RF operations in the vicinity of licensed LTE band challenges the device to support simultaneous radio transmission, which will result in a totally different interworking experience of data off-loading between licensed and unlicensed mobile access. A great many new interference and scheduling challenges are also expected.

By issuing a detailed proposal to the citizens broadband radio service[53] (CBRS) in the 3.5 GHz band, the FCC unveiled a band to investigate innovative spectrum sharing policies. These novel features are demonstrated in the tiered access, interference coordination through spectrum access system (SAS), and dynamic TDD coexistence targets the improvement of spectrum utilization and enhancement of spectral efficiency. It requires more detailed engineering work to support and validate but may eventually lead to an innovated mobile architecture.

In the current device architecture, there are two separated RF FEMs with two sets of device antenna. One is shared by all licensed FDD cellular accesses from 2G/3G to 4G LTE. And, the other is shared by the unlicensed Wi-Fi, Bluetooth, NFC, GPS, and other peripheral radio accesses. To support the CBRS, a third TDD FEM and antenna set might be required to support the dynamic configuration to the licensed and unlicensed user accesses. It not only bridges the dynamic configuration between the licensed and unlicensed network access, but also leads to the coexistence of licensed FDD and TDD within the same mobile device. Numerous challenges must be addressed in such concepts.

13.11   CONCLUSIONS TO THE MOBILE DEVICE EVOLUTION UNDER OPERATOR PERSPECTIVES

To support feature-rich mobile data access, real-time services, audio/video streaming delivery, and IoT service connections, mobile devices hold a unique position to enhance the users' QoE. This understanding directs the network operator's service innovations, technical supports, and architectural developments. For example, to fulfill the vision of LTE advanced, 8×8 MIMO has been specified in the mobile downlink operation in the RAN and 4×4 MIMO has been defined in the mobile uplink access. In the current LTE operations, there are big gaps in targeted performances such that only 2×2 MIMO is implemented in the RAN downlink, and only 1×2 MIMO is carried out in the mobile device uplink access. VoLTE is another example in the LTE evolution. Without VoLTE capability, LTE mobile devices can only offer data access and must rely on 2G or 3G CSFB voice connectivity.

Beyond near-term developments, network operators are looking for long-term opportunities from network evolution. In the past, network configurations have been centered and dominated by the hardware setup. Current views to future network configurations will be highlighted by the virtualized functionalities and intelligent network controls that are characterized by the implementation of SDN. In the view of network operators, continuously optimizing the network configuration for higher efficiency has been a priority in operations. To optimize the consumers' experience is also a priority for network operation. All of these service priorities drive mobile device evolution.

Along with the growth of mobile data and video traffic, additional mobile traffic volumes will come from the operation of IoT over wireless. The IoT industrial M2M communications[54] will have a large portion of rising data traffic including the fixed-to-fixed and fixed-to-mobile device traffic, which might be configured under the licensed/unlicensed data coverage. Beyond the M2M, the efforts and investments to build the infrastructure for V2V communications have seen fast progress. As mobile services shift from data access to video display and mobile TV broadcasting, the balance strategy of mobile access has evolved from licensed LTE to unlicensed data off-loading, and then to the required session continuity in support of the streaming content delivery.

All of the aforementioned evolutions push mobile devices to support the MMMB access capabilities under a new RF spectrum environment. The device access capabilities will not only include more bands and modes through 2G, 3G, and 4G LTE, but also covered by the FDD/TDD coexistence across the licensed and unlicensed spectrums. With SDN virtualization in the network configuration, the trends of SDR lead the mobile device evolution. Under a given device volume and limited battery power supply, a configurable and reconfigurable device FEM becomes the most critical area for technical innovations and developments.

Tuned device RF FEM has already emerged in device antenna development. It enables the same piece of device antenna to cover a wide RF range operation with one band active at a time. It will be evolved to track and compensate for a detuned antenna in a dynamic fashion and also tune the RF performance from a single antenna to a MIMO antenna array.

Fully reconfigurable and integrated device FEMs are the next phase to this evolution. RF CMOS developments have been continuously improving the performances in terms of integrated device FEM as its main target. The RF CMOS architecture covers the RFIC and has been extended to cover MEMS developments in the RF switches and PAM developments.

Simultaneous radio operation will be a key milestone to raise the device efficiency in support of higher-order MIMO and multiradio access. It is challenged by the coordination of device RF interference and effective battery power management. Phased progress characterizes device development from the near-term to the long-term evolution.

In the current device FEM architecture, the RF transceiver function is covered by analog operations where the RF signals are filtered, amplified, and/or modulated/demodulated by the analog circuits in the device FEM. SDR implementation is purely in the digital domain. The demarcation between analog and digital operations is at the ADC/DAC conversion. Raising the ADC RF sampling rate and maintaining the ADC performance at a high-bit resolution and large dynamic range still fundamentally challenges SDR implementation without a trade-off to device battery life. This becomes one of the expectations of the network operators. Working with the spectrum regulatory administration and the vendor community, we are at the beginning of this long-term journey. The challenges are highly technical, but the innovations are user QoE directed. All of the technical innovations and advancements are driven by market demand.

REFERENCES

1. www.keysight.com/upload/cmc_upload/All/23Jan14WebCastSlides.pdf; www.gsacom. com, updated February 5, 2014.
2. http://www.worldtimezone.com/4g.html, updated February 12, 2014.
3. http://www.cisco.com/c/en/us/solutions/collateral/service-provider/visual-networking-index-vni/white_paper_c11-520862.pdf.
4. http://www.youtube.com/watch?v=tIv0jhMa64E.
5. http://www.4gamericas.org/documents/2013_4G%20Americas%20Meeting%20 the%201000x%20Challenge%2010%204%2013_FINAL.pdf.
6. http://stakeholders.ofcom.org.uk/consultations/ddr/statement/.
7. http://www.qualcomm.com/research/projects/lte-advanced/asa.
8. http://www.gsma.com/spectrum/wp-content/uploads/2013/04/GSMA-Policy-Position-on-LSA-ASA.pdf.
9. http://www.fiercewireless.com/tech/story/atts-rinne-using-son-helps-improve-through-put-and-reduce-dropped-calls/2012-10-30.
10. http://www.multichannel.com/distribution/att-u-verse-expands-live-tv-lineup-tablets-smartphones/148633.
11. http://www.phonearena.com/news/AT-T-uses-carrier-aggregation-to-quietly-bring-LTE-Advanced-to-the-Windy-City_id53600.
12. http://www.qualcomm.com/solutions/wireless-networks/technologies/carrier-aggregation.
13. http://www.computerworld.com/s/article/9248366/AT_T_to_activate_HD_Voice_over_4G_LTE_in_four_states_on_May_23?source=rss_latest_content.
14. Camarillo, G. and García-Martín, M.-A., *The 3G IP Multimedia Subsystem: Merging the Internet and the Cellular Worlds*, John Wiley & Sons, 2008, ISBN-13: 978-0-470-69513-5.
15. Narang, N. and Kasera, S., Architecture, Protocols, and Procedures and Services, In: *2G Mobile Networks: GSM and HSCSD*, Tata McGraw-Hill Publishing Company Limited, 2007, ISBN 0-07-062106-3.
16. Korhonen, J., *Introduction to 3G Mobile Communications*, 2nd edn, Artech House, Inc., 2003, ISBN 1-58053-507-0.
17. Kyriazakos, S., Soldatos, I., and Karetsos, G., *4G Mobile & Wireless Communications Technologies*, CTiF (Aalborg University), Athens Information Technology, Technology Research Center of Thessaly, River Publishers, 2008, ISBN: 9788792329028.
18. http://en.wikipedia.org/wiki/Orthogonal_frequency-division_multiplexing.
19. Wang, Z. and Giannakis, G.B., Wireless multicarrier communications: Where Fourier meets Shannon, *IEEE Signal Processing Magazine*, 17(3), 29–48, 2000.
20. http://en.wikipedia.org/wiki/Single-carrier_FDMA.
21. http://people.rit.edu/grteee/media/IST03.pdf.
22. http://en.wikipedia.org/wiki/MIMO.
23. http://appleinsider.com/articles/14/02/24/broadcoms-new-5g-wi-fi-chip-with-80211ac-2×2-mimo-could-be-tapped-for-future-iphone.
24. SkyCross presentation: *Advances in MIMO for LTE-A to IWPC Workshop*, San Jose, CA, March 20, 2014.
25. http://www.anandtech.com/show/7921/qualcomm-announces-mumimo-80211ac-family-increasing-the-efficiency-of-80211ac-networks.
26. http://en.wikipedia.org/wiki/LTE_Advanced.
27. http://www.4gamericas.org/index.cfm?fuseaction=page&sectionid=352.
28. http://reviews.cnet.com/8301-13970_7-57619375-78/at-t-expands-international-lte-roaming-to-13-more-countries.

29. http://www.samsung.com/global/business/business-images/resource/white-paper/2013/02/eMBMS-with-Samsung-0.pdf.

30. http://www.ericsson.com/res/docs/whitepapers/wp-lte-broadcast.pdf.

31. http://www.att.com/gen/press-room?pid=25262&cdvn=news&newsarticleid=37430&mapcode=consumer|enterprise.

32. http://business.time.com/2014/01/07/your-car-is-about-to-get-smarter-than-you-are/, January 7, 2014.

33. http://www.att.com/gen/press-room?pid=25274&cdvn=news&newsarticleid=37439&mapcode=.

34. http://downloads.lightreading.com/wplib/cisco/HR_Cisco_LTE_Signaling_WP_Final.pdf.

35. http://ieeexplore.ieee.org/xpl/login.jsp?tp=&arnumber=6848317&url=http%3A%2F%2Fieeexplore.ieee.org%2Fiel7%2F6842514%2F6847849%2F06848317.pdf%3Farnumber%3D6848317.

36. Tunable multiband power amplifier using thin-film BST varactors for 4G handheld applications, ISBN: 978-1-4244-4725-1.

37. https://www.nttdocomo.co.jp/english/info/media_center/pr/2010/001466.html.

38. http://en.wikipedia.org/wiki/Lithium-ion_battery.

39. Nokia Siemens Networks, More battery life with LTE connected state DRX. http://networks.nokia.com/system/files/document/more_battery_life_with-lte_connected_drx_0.pdf.

40. http://www.stuff.co.za/startup-eta-devices-may-soon-provide-tech-that-halves-smart-phone-power-consumption/.

41. Nujira Ltd., Envelope tracking: Unlocking the potential of CMOS PAs in 4G smart-phones, http://www.nujira.com/white-paper--et---unlocking-the-potential-of-cmos-pas-in-4g-smartphones-i-369.php, February 2013.

42. http://electronicdesign.com/communications/understanding-mipi-alliance-interface-specifications.

43. http://www.nvidia.com/content/PDF/tegra_white_papers/Benefits-of-Multi-core-CPUs-in-Mobile-Devices_Ver1.2.pdf.

44. http://www.nvidia.com/content/PDF/tegra_white_papers/tegra-whitepaper-0911a.pdf.

45. http://www.computerworld.com/article/2866924/samsungs-galaxy-a7-packs-bigger-screen-more-processing-power.html.

46. Fingas, J., Windows Phone 8 to support multi-core CPUs, HD resolutions, SD cards and NFC, Engadget, AOL Inc., June 20, 2012, Retrieved November 26, 2012. http://www.engadget.com/2012/06/20/windows-phone-8-to-support-multi-core-cpus-hd-resolutions/.

47. http://lteportal.com/Files/MarketSpace/Download/447_1-14nomor.pdf.

48. http://www.bell-labs.com/user/supratim/pubs/eICIC-Public.pdf.

49. http://books.google.com/books?id=_B_nAgAAQBAJ&pg=PA133&lpg=PA133&dq=further+enhanced+Inter-cell+interference+coordination+(FeICIC)&source=bl&ots=WXihsYS-ld&sig=FVBOKvd4GDUDOTpzaRa6BU7SwzI&hl=en&sa=X&ei=l8hWU-q0O4bA2QW5loHIAw&ved=0CE4Q6AEwCA#v=onepage&q=further%20enhanced%20Inter-cell%20interference%20coordination%20(FeICIC)&f=false.

50. http://www.radio-electronics.com/info/cellulartelecomms/lte-long-term-evolution/4g-lte-advanced-comp-coordinated-multipoint.php.

51. http://www.tuhh.de/ffv/WS/workshop_2014_1/ffv-2014-christian-hoymann.pdf.

52. http://www2.ic.uff.br/~ejulio/doutorado/artigos/06272421.pdf.

53. http://www.fcc.gov/blog/35-ghz-new-ideas-innovation-band.

54. http://52ebad10ee97eea25d5e-d7d40819259e7d3022d9ad53e3694148.r84.cf3.rackcdn.com/UK_SIW_LTE%20and%20M2M%20converging%20paths%20_WP.pdf.

55. http://www.home.agilent.com/upload/cmc_upload/All/23Jan14WebcastSlides.pdf.
56. http://www.cedmagazine.com/news/2013/06/mobile-video-consumption-growing-at-60-percent/year.
57. Carson, P. and Brown, S., Less is more: The new mobile RF front-end, *Microwave Journal*, 56(6), 24–34, 2013.
58. Lou, Y., Evolved Mobile Broadband Services & Devices, Presentation to *IWPC Workshop on Optimizing Advance Smartphone and Tablets Architecture*, March 19, 2014, http://www.iwpc.org/ResearchLibrary.aspx?ArchiveID=210&Display=doc#.

14 Testing Wireless Devices in Manufacturing

Rob Brownstein and Minh-Chau Huynh

CONTENTS

14.1 INTRODUCTION

Only a decade and a half into the twenty-first century, wireless devices abound. From wireless car keys that allow you to open a car door and start the engine without leaving your pocket to smartphones that let you make calls, surf the Internet, take and transfer photos and videos, and pay for items electronically, people have essentially tether-free access to each other and a multitude of services.

Today's smartphone eclipses in complexity computing and communications devices that were densely packed with components, took up cubic yards of space, and cost tens of thousands of dollars. And, this, by a device one holds in a hand that contains computing, networking, and wireless communications subsystems and costs only a few hundred dollars to manufacture. It is no wonder, then, that millions of these devices are produced and sold each month. The amazing part is that despite the complexities, such devices can be mass produced in huge volumes, with high yields. Furthermore, manufacturing testing has become so efficient and fast that essentially every device can be tested to help ensure only a tiny fraction of faulty devices end up in the hands of customers.

Because such wireless devices are combinations of computers, network clients, and wireless transceivers, testing them during manufacturing entails making sure that they work as designed. Testing costs are a combination of capital equipment, facilities, power, and labor costs. There is a direct correlation between testing cost and testing time. With very tight testing cost budgets due to price competition and thinning margins, there is huge pressure on companies that design and sell test solutions to wireless device manufacturers to decrease testing cost per device even as the number of functions to be tested increases significantly.

As with any computer, sophisticated wireless devices such as smartphones and cellular-enabled tablets have central processing units (CPUs), operating systems, storage, and I/O. For self-contained applications, such as music playback, taking photos or videos, or playing games, there are manufacturing tests for ensuring that a device boots up properly and that all related subsystems, I/O, and transducers (e.g., touch screen, LCD display, speaker, microphone) are all working as designed.

When these devices must upload or download data to other devices, they do so wirelessly. Typically, a smartphone or tablet will have Bluetooth for hands-free calling in an automobile. They will also have Wi-Fi for using the Internet directly, rather than through the metered cellular service for Internet access. This, of course, requires being close enough to a Wi-Fi hotspot to accommodate such use. Cellular system access typically requires one or more radio-access technologies, such as 3G and 4G (Third Generation Partnership Project Long Term Evolution [3GPP LTE]). These will have a wireless subsystem associated with them, too, such as a transmitter and receiver. Thus, testing of the wireless functions of a device will involve tests different from those for the computing subsystems.

This chapter focuses on the testing of wireless functions during manufacturing. A device, for example, that features 3G and 4G (3GPP LTE), 802.11n Wi-Fi, and Bluetooth technologies will be tested to ensure that each one is still operating properly after manufacture. The assumption is that the design has been verified, earlier, and that a validated design has moved to production. In other words, manufacturing test is not used to verify a design, but, rather, to identify any manufacturing defects that may have occurred and which now render the device's wireless facilities either faulty or out of specification.

Several technologies are common to cellular and so-called connectivity wireless technologies, such as Wi-Fi. These common denominators include orthogonal frequency division multiplexing (OFDM), quadrature amplitude modulation (QAM), and multiple-input multiple-output (MIMO). In addition, these wireless technologies require antennas. To fully test a device that replicates real-world operation, the fully assembled device should be linked to a tester in the same way it would to an access point or base station, using over-the-air (OTA) signals. Today, this real-world test typically occurs last. Before that, the partially assembled device has been tested using non-linked processes for virtually all the specifications and functions; and it is the partially assembled device that has passed these tests prior to the final fully assembled device test. There are many reasons for why this process is in place, and they will be described. Recently, innovative technologies have been developed that make testing fully assembled devices using OTA signal transfer more practical than before. These, too, will be described and explored.

14.2   Wi-Fi AND 4G (3GPP LTE)

It is interesting to explore the evolution of cellular from its original analog AMPS roots to its current digital state of the art. It is also interesting to do the same for Wi-Fi, which is the generic term for the IEEE 802.11 standard and all its subsequent amended versions (e.g., 802.11a, b, g, n, etc.). However, several volumes could be written on these evolutionary changes. Instead, let's briefly explore two key wireless standards. The IEEE 802.11n standard is currently the most widely used; and the 3GPP LTE standard, herein called "4G LTE," is the current state of the art for cellular.

14.2.1   IEEE 802.11n

Prior to amendments underlying IEEE 802.11n, wired Ethernet speeds were shooting past those of wireless Ethernet. In order of their adoption, IEEE 802.11b, a, and g established a beachhead for wireless Ethernet, but 802.11n really set the stage for significant performance improvements.

Several changes underpin 802.11n's performance enhancements. By adopting OFDM, higher data throughput compared with the previous a, b, and g versions of the standard is achieved by having a networking system all using the 802.11n standard. A new PHY layer convergence protocol (PLCP) supporting the 802.11n standard with no legacy provisions (called Greenfield) enables highest data throughputs. A PLCP called "mixed mode" uses a preamble consistent with 802.11 a and g but incorporates some performance enhancements, such as a MIMO training sequence. This allows a performance improvement on a networking system that includes clients with 802.11 a or g technologies. The third PLCP, legacy mode, simply duplicates 802.11 a or g modes. Which PLCP modes are brought to bear is determined by the makeup of the networking system and its clients, and the mode being used determines the characteristics of the PHY layer signal formats being exchanged. Note, also, that the radio frequency (RF) bands prescribed cover the 5 GHz band of 802.11 a and the 2.4 GHz band of 802.11 g.

Another performance booster in 802.11n is the use of MIMO. Compared to 802.11 a and g, 802.11n can improve range using spatial diversity and can improve data throughput using spatial multiplexing. Both improvements are a result of using MIMO technology. With 802.11n, up to four spatial streams can be used, which would require a device having four transmitters, four receivers, and four antennas (a $4 \times 4 \times 4$) engaging with an access point having the same capabilities.

By supporting 20 or 40 MHz channel widths, one can trade-off the number of channels to increase data throughput. Also, IEEE 802.11n supports beamforming when using two or more transmission subsystems and antennas. Beamforming optimizes the path between the wireless device and the access point, reducing the effect of multipath destructive interference.

The IEEE 802.11n standard document, known as Part 11: Wireless LAN Medium Access Control (MAC) and Physical Layer (PHY) Specifications; Amendment 5: Enhancements for Higher Throughputs, is more than 500 pages long and covers just the Wireless MAC and Physical Layer (PHY) Specifications. The key portions of the standard that drive the manufacturing testing of 802.11n devices are: 20.3.21 PMD transmit specification and all its subsections, and 20.3.22 HT PMD receiver specification and all its subsections.

For example, in 20.3.21.1, Transmit Spectrum Mask, which underpins a transmitter (TX) test, the limits are established for devices using 20 MHz and 40 MHz channels. In the first case, the standard states:

When transmitting in a 20 MHz channel, the transmitted spectrum shall have a 0 dBr (dB relative to the maximum spectral density of the signal) bandwidth not exceeding 18 MHz, −20 dBr at 11 MHz frequency offset, −28 dBr at 20 MHz frequency offset, and the maximum of −45 dBr and −53 dBm/MHz at 30 MHz frequency offset and above. The transmitted spectral density of the transmitted signal shall fall within the spectral mask, as shown in Figure 14.1. The measurements shall be made using a 100 kHz resolution bandwidth and a 30 kHz video bandwidth.

And Figure 14.1 depicts the spectral mask for 20 MHz channel testing.

The document goes on to state that for 40 MHz channels,

When transmitting in a 40 MHz channel, the transmitted spectrum shall have a 0 dBr bandwidth not exceeding 38 MHz, −20 dBr at 21 MHz frequency offset, −28 dBr at 40 MHz offset, and the maximum of −45 dBr and −56 dBm/MHz at 60 MHz frequency offset and above. The transmitted spectral density of the transmitted signal shall fall within the spectral mask, as shown in Figure 14.2.

Figure 14.2 depicts the spectral mask for 40 MHz channel testing.

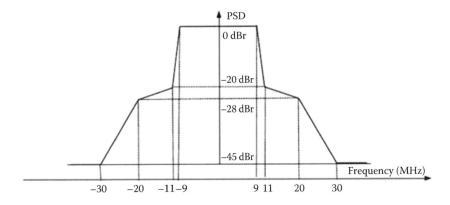

FIGURE 14.1   Transmit spectral mask for 20 MHz transmission.

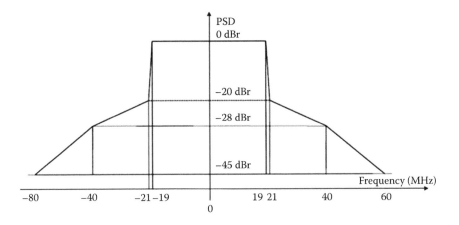

FIGURE 14.2   Transmit spectral mask for 40 MHz channel.

The 802.11n standard also establishes spectral flatness limiting the delta in average power for the subcarriers for devices that use 20 MHz and/or 40 MHz channels. In the 20 MHz channel, there are two specifications. In subcarriers that correspond to constellation indices −16 to −1 and +1 to +16, the average power cannot deviate more than ±2 dB. For subcarriers corresponding to constellation indices −28 to −17 and +17 to +28, average power cannot deviate more than +2/−4 dB from the average power measurements of the first set of subcarriers. Maximum transmit power is determined by the regulations in each country and the standard provides some tables for the United States, Europe, and Japan in the annex section (Annex I) of the standard.

Center frequency tolerances are established for devices operating in the 2.4 GHz spectrum and for those operating in the 5 GHz spectrum. Interestingly, only ±20 ppm is allowed for the higher frequency spectrum versus ±25 ppm for the 2.4 GHz one. The numbers are identical for signal clock frequency tolerances.

The 802.11n standard specifies the limit of error in constellations. Ideally, the subcarriers would have perfectly orthogonal positions relative to each other. In reality, of course, they do not. The limit to how much they may deviate is related to constellation size and coding rate as, for example, BPSK with coding rate ½ is limited to −5 dB of constellation error, whereas 64 QAM with coding rate of 5/6 is limited to −29 dB of constellation error.

Transmit modulation quality is determined by error-vector magnitude (EVM) metrics. Here is how the standard describes this test procedure:

> The transmit modulation accuracy test shall be performed by instrumentation capable of converting the transmitted signals into a streams of complex samples at 40 Msample/s or more, with sufficient accuracy in terms of I/Q arm amplitude and phase balance, dc offsets, phase noise, and analog-to-digital quantization noise. Each transmit chain is connected directly through a cable to the setup input port. A possible embodiment of such a setup is converting the signals to a low intermediate frequency with a microwave synthesizer, sampling the signal with a digital oscilloscope, and decomposing it digitally into quadrature components.

The standard goes on to describe a step-by-step procedure or its equivalent, for measuring EVM.

For receiver (RX) specifications, one refers to 20.3.22. The first specification is for receiver minimum input sensitivity and it is determined by a packet-error rate (PER) limit of 10%. Minimum sensitivity for 20 MHz channels using 64 QAM modulation at a rate of 5/6 is −64 dBm, and for a 40 MHz channel, −61 dBm.

Note, also, that the standard specifies the limits for adjacent and nonadjacent channel rejection that are allowed. Rejection decreases with the higher throughput QAM modulation compared with binary phase-shift keying (BPSK) and quadrature phase-shift keying (QPSK). Rejection is determined by setting a channel's level at 3 dB above the minimum sensitivity value and then increasing the power of the other channel's signal until a 10% PER occurs in the first channel.

In addition to a minimum sensitivity level, the specification also sets a maximum level above which saturation could occur causing the PER to exceed 10%.

14.2.2   4G LTE

Wi-Fi from its inception was, like the Internet, meant to be part of a global digital network. As such, the IEEE 802.11 standards have been adopted virtually everywhere in the world. In fact, using voice-over-IP, one can even use Wi-Fi to facilitate global telephone communications using applications such as Skype. Of course, doing so requires one to be able to link to a Wi-Fi hotspot.

By design, cellular services offer a much broader service area than Wi-Fi. Elaborate systems of cellular base stations connected to wired telephone infrastructure allow users to be connected wherever cellular service is accessible. The first cellular standards, such as AMPS, were analog systems. Later, the first digital telephony standards and services emerged. These were termed second-generation systems, or 2G. More sophisticated digital services, including both telephone and data, were the hot buttons of third-generation standards, also known as 3G. And, today, we have 3GPP LTE standard, which much like 802.11n, has made cellular data services more robust and higher performance.

The standard that underlies 4G LTE manufacturing testing is 3GPP TS 36.521-1. Like IEEE 802.11, it has undergone a succession of versions, each amended with new and enhanced elements. LTE, for example, was introduced initially in Release 8. It introduced a radically different and new radio interface and network. Downlink speed maximum was 300 Mbps and maximum speed for uplink was 75 Mbps. Signal bandwidths of 1.4, 3, 5, 10, 15, or 20 MHz were supported allowing for different implementation scenarios.

One of the fundamental decisions was to use different access schemes for download and upload. Downlink uses orthogonal frequency domain multiple access (OFDMA), whereas uplink uses single-carrier frequency domain multiple access (SC-FDMA). The benefit of SC-FDMA uplink access is better energy efficiency because of lower average output power of a device's transmitter power amplifier. As with 802.11n, LTE takes advantage of MIMO technology for both spatial diversity (e.g., range extension) and spatial multiplexing (e.g., higher throughput). In LTE, the centralized control of earlier networks (e.g., GSM BSC or UMTS RNC) was eliminated in favor of a flat distributed architecture (see Figure 14.3).

Further standard releases have followed Release 8 up to and including Release 12, which was frozen in 2014. As a result, the initial downlink and uplink maximum speeds have increased to 3 and 1.5 Gbps, respectively (Release 10). In addition, a feature called "carrier aggregation" has been added that allows up to five carriers to be aggregated achieving an effective bandwidth of 100 MHz (Release 10).

As with IEEE 802.11n, the PHY layer specifications of 4G LTE drive the PHY layer manufacturing test of devices that employ it. Thus, for example, physical layer testing focuses on the lowest layer of the air interface. It seeks to determine conformance with the key parameters essential to the successful transmission of a signal over the air. Transmit power, the quality of the TX waveform, the accuracy of the TX frequency, are all key to a mobile station's performance. On the receive (RX) side, the ability of the mobile to successfully decode the received signal at the lowest and highest signal levels defines its successful operation in the network.

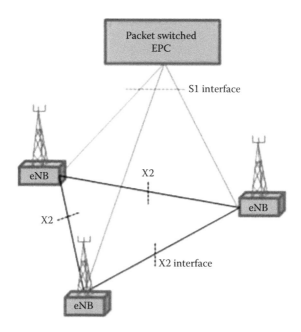

FIGURE 14.3   4G LTE replaced the centralized architecture of previous standards with a flat, distributed architecture.

As such, the pertinent specifications are TX power, EVM, frequency error, adjacent channel leakage ratio (ACLR), spectrum emission mask, occupied bandwidth, carrier leakage, TX time mask, and in-band emissions from non-allocated resource blocks. In many ways, the tests at PHY layer of 4G LTE signals are very similar to test for the analogous characteristics of an 802.11n signal. To be sure, because of the complexity of 4G LTE relative to 802.11, there are many tables in the specifications covering power metrics where there is no carrier aggregation or UL MIMO compared to cases where there are. The tables that are most relevant to PHY layer testing are maximum output power, which corresponds to Section 6.2.2 of the specifications, and minimum output power, which corresponds to Section 6.3.2. As noted, later, a device that passes those two TX power tests can be considered compliant with the other tables and limitations because these values are controlled by software and subsystems in the device that are unlikely to be faulty if the basic maximum and minimum power tests indicate correct operation.

With regard to EVM, the 4G LTE specifications are quite explicit. For example, the specification states:

> The RMS average of the basic EVM measurements for 10 sub-frames excluding any transient period for the average EVM case, and 60 sub-frames excluding any transient period for the reference signal EVM case, for the different modulations schemes shall not exceed the values specified in Table 6.5.2.1.1-1 for the parameters defined in Table 6.5.2.1.1-2. For EVM evaluation purposes, [all PRACH preamble formats 0-4 and] all PUCCH formats 1, 1a, 1b, 2, 2a and 2b are considered to have the same EVM requirement as QPSK modulated.

For QPSK and BPSK, the average EVM level cannot exceed 17.5%; and for 16 QAM, it is a more stringent 12.5%.

Frequency error over a period of one time slot (0.5 ms) is limited to ±0.1 ppm compared to the carrier frequency received from the network node.

There are related specifications that are essentially looking for the same potential problem—adjacent channel interference. The adjacent channel leakage ratio looks at the amount of power found in an adjacent channel due to a signal in the other channel. The 4G LTE standard specifies two different cases, one where the channels are adjacent Evolved Universal Terrestrial Radio Access (E-UTRA) channels and one where they are UTRA channels. Most LTE conformance tests for out-of-band emissions are similar to those used for W-CDMA. These are consistent with pre-LTE 3GPP releases, where UTRA channels are employed, where a root-raised cosine filter is used for making transmitter measurements. LTE, on the other hand, is not constrained to root-raised cosine filters in its E-UTRA channels, which result in better ACLR characteristics. Hence, we have two different specifications.

Essentially, the ratio is of the filtered mean power of the signal centered in the assigned channel to that measured centered in the adjacent channel. Another metric—occupied bandwidth—looks at the bandwidth of a signal in an assigned channel and compares the bandwidth in which 99% of the total integrated power resides to that of the channel. It is checking to make sure that the signal occupies less than the channel bandwidth. And the spectrum emission mask specifies the roll-off of power on either side of a signal. It is analogous to the spectral mask test used in testing Wi-Fi. If a device's TX signal passes the mask test, it is unlikely to fail the ACLR test.

Carrier leakage refers to carrier signal present in the signal output. In a properly matched transmitter, the IQ modulation suppresses the carrier energy to a point where it is attenuated far below the normal signal's energy. The carrier leakage specification for 4G LTE specifies different leakage levels at different signal power levels.

The TX time mask test looks at the signal in time, verifying that the transmitter power amplifier is turning on and off at the correct time without producing any extraneous signals. Since LTE signals are shared both in frequency and time, being accurate in the time domain is just as important as being accurate in the frequency domain.

The in-band emissions test for non-allocated resource blocks (RBs) is meant to ensure that RB subcarriers in a non-allocated RB are not emitting more than a very limited energy, in band. It is a ratio of the power of non-allocated RBs to allocated RBs over a time slot and averaged across 12 subcarriers.

These 4G LTE TX specifications (e.g., uplink) are the basis for the test flow in LTE TX manufacturing testing. Similarly, the following 4G LTE RX specifications drive the test flow in LTE RX manufacturing testing.

The standard states that unless otherwise stated, the receiver characteristics are specified at the antenna connector(s) of the user equipment (UE). "For UE(s) with an integral antenna only, a reference antenna(s) with a gain of 0 dBi is assumed for each antenna port(s). UE with an integral antenna(s) may be taken into account by converting these power levels into field strength requirements, assuming a 0 dBi gain antenna. For UEs with more than one receiver antenna connector, identical interfering signals shall be applied to each receiver antenna port if more than one of these is used (diversity)." Thus, the signals specified are either applied directly to an antenna connector, or, where an antenna is integral to the device, it is assumed to be a 0 dBi antenna, that is, having no directional gain. In addition, the specifications assume a MIMO baseline of 2. That is, the device is assumed to have two RX signal ports.

With Wi-Fi, RX tests, by and large, are performed against a PER specification. For example, minimum and maximum receiver (RX) power sensitivity is determined when PER exceeds 10%. With 4G LTE, RX test specifications are based on a reference sensitivity power level. Maximum and minimum power levels in 4G LTE are not determined by error rate, irrespective of data rate, but rather by the power levels at which a device can achieve 95% or better of the maximum throughput. The standard contains different tables for cases where there is no carrier aggregation and cases where there is carrier aggregation.

To be sure, the specifications contain a lot of limits, tables, and the like with regard to RX performance. These include reference sensitivity power level, maximum input level, UE maximum input level with carrier aggregation, adjacent channel selectivity, blocking characteristics, and spurious emissions. However, assuming the device is properly designed and validated, these measurements are not likely to fail manufacturing testing so long as maximum and minimum power level tests have been passed. That allows for limiting test steps without risking false negatives.

14.3 MANUFACTURING TEST METHODS AND APPARATUS

Unlike the equipment used for device design verification, including signal generators, spectrum analyzers, and multifaceted laboratory instrumentation, the equipment used for manufacturing test is intended for accuracy, speed, and test integrity. The first types of manufacturing test solutions designed for high-volume testing of wireless devices (e.g., Wi-Fi NICs) were so-called single-box testers containing vector signal analysis, vector signal generation, and control subsystems used in conjunction with computers. These boxes tended to be purpose-built machines for testing Wi-Fi or Bluetooth or other high-volume devices. It was the advent of smartphones containing cellular, Wi-Fi, Bluetooth, and other radios that led to manufacturing testers that could test multiple technologies.

Typically, partially assembled devices are tested for wireless capability by conducted rather than radiated signal means. That is, during RX testing of the devices, test signals from the computer-controlled vector signal generation subsystem are conveyed to the device or devices by coaxial cables connected at one end to the test system and at the other end to the device's antenna connector. It is very easy to calibrate the cable loss and, therefore, know to a high degree of accuracy the signal power being applied to the device's antenna connector. Thus, one can vary power level, modulation type, and frequency while monitoring PER to determine, say, RX sensitivity.

Similarly, during TX testing, a device's antenna connector is used to convey signals from the device to the tester's vector analysis subsystem. Again, by knowing precise cable loss, one can accurately determine the TX output power at a device's antenna connector by measuring the input power at the tester's input port connector. Here, again, one can control the device's power, frequency, and data rates while measuring spectral mask, maximum and minimum power, and EVM with the tester's vector analysis subsystem.

During device design verification, virtually all of a standard's specifications are tested to make sure the device has been designed properly. The benefit of this is that the manufacturing test steps and flow can be a subset of verification tests that determine whether any manufacturing defects have rendered a device faulty or out of specification without having to reaffirm that all the specification limitations are still met. Thus, testing can be confined to the PHY layer subsystems without undue fear that some other failure will go undetected.

The "art" in developing manufacturing test solutions is not necessarily in the design of the vector signal generation or analysis subsystems, or the control systems. The art is in developing the test control interface that enables the tester to control the device through the device's chipset. Even where two different devices have the same chipset, the test programs for each device may be significantly different because of other differences in features and technologies.

In looking at a wireless test regimen, one must include the time spent connecting and disconnecting devices to the test system, the time it takes to boot up the device, and the time it takes to conduct the actual TX and RX tests. Since 2005, with Wi-Fi and Bluetooth, test engineers have reduced the time it takes to conduct the actual TX and RX tests to a point of diminishing return. Further reduction in test time and costs has been gained by innovations in device-under-test (DUT) handling as well as reduction of DUT–tester communications so that a larger proportion of the time spent sending signals back and forth is test, rather than control, related.

As a single device's overall test time has been reduced to diminishing return dimensions, further time savings have been gained by testing devices concurrently. If one can test, say, four devices in a time that is only 20% longer than testing a single device, the overall test time saving is huge—nearly 75%. If one can test a single device's multiple technologies in a quasi-concurrent fashion, one can again achieve significant time savings. Where the objective is to test millions of devices per month that test time saving can be the difference between hitting a market window or not. With intense price competition on many wireless devices, the faster test times can translate into lower test costs and preserve margins.

One thing, however, should be apparent. If most of the manufacturing tests are done on a partially assembled DUT, and signals are conveyed using conductive means rather than radiated means, once the device is fully assembled, including its antenna(s), there are still possibilities for manufacturing defects. Thus, it is not uncommon for the assembled devices to be final tested in a way that mimics the way they will actually be used. That is, the devices are linked to an access point using that technology's linking processes, and the signals are conveyed between tester and DUT using radiated conveyance rather than conducted means.

Testing of the fully assembled device using OTA signals certainly replicates real-world operation. And if the objective is simply to determine that after final assembly, when the antenna(s) is (are) connected, the device still works, then current OTA testing environments are probably adequate. The problem, however, with current OTA testing environments is that they do not allow the same level of precise testing that conductive testing affords. As already noted, one can easily determine cable loss. Furthermore, one can create sufficient isolation to ensure that only the signal of interest is being tested. In an OTA environment, as anyone who has ever dropped a call while walking can attest, path losses are very position and orientation dependent and not very predictable.

If there was a way to make the OTA environment more precise and predictable, where one could control path loss to a point where DUT position was no longer critical and a reliable loss metric could be applied to both path directions (e.g., from DUT to tester and from tester to DUT), it would then be possible to do many of the tests now confined to conductive conveyance by using OTA conveyance. And with some of the newer device designs featuring "tuned" antennas for optimizing signal strength and reducing battery power drain, it would be possible to ensure that these tunable antennas were working properly after final assembly. Recent breakthroughs in OTA test infrastructure appear poised to do exactly this.

14.4   WIRELESS TESTING OF WIRELESS DEVICES

The idea of using OTA signals to test wireless devices is not new. However, practical sizes and predictable path losses have been serious obstacles. However, a recent breakthrough in OTA technology is paving the way for using OTA testing for MIMO devices and concurrent multiple DUT testing.

One patented technology (US 8,917,761), for example, makes use of an array of antennas within an enclosed, shielded chamber designed to encourage multipath signal reception. The enclosure provides isolation but without absorbing material inside, the interior of the enclosure presents a rich multipath environment. By varying the phase differences and magnitudes of signals applied to individual array antennas, one reduces the contribution of cross-coupled signals and normalizes the resultant. In other words, the matrix values of h_{ij}, where $i = j$, are normalized by controlling magnitudes, whereas those for $i \neq j$ are minimized using selective phase differences (see Figure 14.4a, the patent's Figure 4).

In the end, one achieves a channel characteristic similar to conductive conveyance. The upshot is that one can convey signals between DUT and tester, using radiated signals, while achieving the predictability and repeatability of using conductive means for conveyance. In addition, measurements using this approach correlate well with those done using far-field measurement capabilities.

Figure 14.4b shows the patent's Figure 5 where one can see in that particular embodiment the DUT, and the array antennas are enclosed in the chamber and the rest of the system is external. A closed-loop control uses measured information to affect the settings of the signal phase controllers on each of the chamber's array antennas.

One of the immediate advantages of this example is that chamber size can be markedly reduced allowing more test setups to be placed in the same floor space. Furthermore, this example can find or adapt the channel environment inside the enclosure to a device's location and orientation to satisfy the purpose of the test. A conventional enclosure does the opposite, in other words, a manual search in DUT position and orientation to find the appropriate channel needed for each frequency or frequency region. The position and orientation is frequency dependent. Sometimes it may not be possible to find the needed channel using the conventional method.

Using this patented technology, one can find an optimal channel that suits the purpose of a test for each frequency, and do so using a single DUT location and orientation.

Earlier OTA test environments, for factory testing in small metallic enclosures, used movable antennas and precise positioning screws that could be adjusted to make path loss more manageable, but the time required to adjust such setups, and the need to readjust each time a new device was placed inside, makes them less practical for large-scale manufacturing testing where accuracy and overall speed prevail.

The invention described in the patent has been reduced to practice and is being productized. Findings, so far, bear out that the path loss after magnitude and phase delay compensations are consistent with using a coaxial cable. The OTA testing system prototype showed that a DUT with two antennas (e.g., a MIMO 2 × 2 device) could also be tested under optimal MIMO performance conditions where all receivers see the same level of signal strength. It was not intended for testing MIMO devices in a fading environment. In addition, the prototype allowed testing of multiple MIMO streams in parallel, thus speeding up testing time. Using conventional methods, it is often difficult or impossible to find a position and orientation of a DUT with two antennas that satisfies the needed channel characteristics, as in this example, where the cross-coupled signals are minimized.

Additional testing showed that one could stack multiple devices in the chamber and establish predictable and repeatable path loss conditions between the tester and DUTs as if connecting a vector signal generated source signal, through a splitter, and routed conductively to those multiple devices.

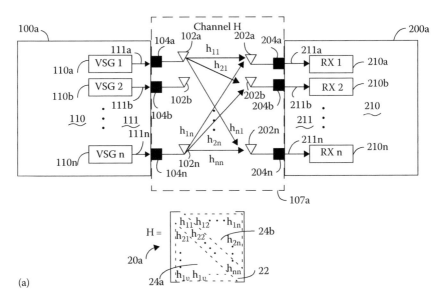

(a)

FIGURE 14.4 Ordinarily, radiated signals from multiple antennas to multiple antennas, as shown, will produce transmitted components of each transmitting antenna at each receiving antenna. By using an array of antenna elements, each having separately controlled phase shifts, one can reduce the components of h=ij where I does not equal j creating an environment where path loss is both predictable and symmetrical. *(Continued)*

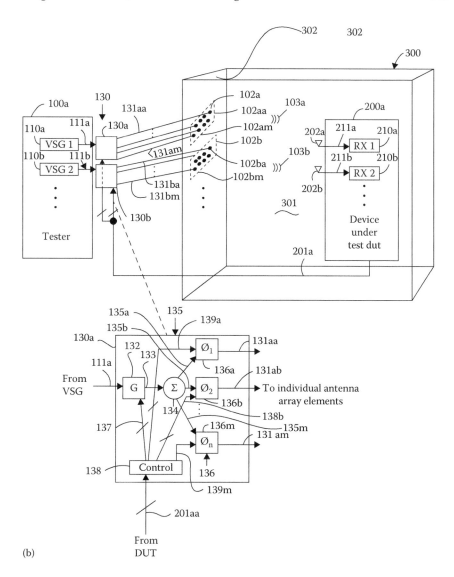

(b)

FIGURE 14.4 (*Continued*) Ordinarily, radiated signals from multiple antennas to multiple antennas, as shown, will produce transmitted components of each transmitting antenna at each receiving antenna. By using an array of antenna elements, each having separately controlled phase shifts, one can reduce the components of $h = ij$ where I does not equal j creating an environment where path loss is both predictable and symmetrical.

14.5 IMPLICATIONS

An OTA testing system that can quickly and easily compensate for multipath signals at a device's antenna(s), to create a reliable and repeatable path loss, can be used as an alternative for contemporary testing using multiple ports and conductive means of conveyance. What that means is that some of the tests currently being done conductively can be shifted to OTA wireless test environments. Rather than having conductive tests on partially assembled devices, and wireless testing of fully assembled devices, one can envision a case where a test flow in the OTA test system can do it all.

One of the advantages of shifting more conductive tests over to OTA tests is being able to test the fully assembled device. In addition, there would be no cabling and uncabling handling times or the prospects of damaging a device's antenna connector(s).

One unique test that an OTA test system can provide is antenna tuning. Doing it requires that a device's antenna is connected, thus precluding a conductive test alternative. One of the unsung areas of device innovations is advances in antenna technology. These advances can take advantage of OTA testing capabilities.

14.6   OTA METHOD FOR ANTENNA TUNING

When a cellular phone was essentially a wireless telephony device, designed to work in a single allocated spectrum, the antenna could be designed as a monoband monopole. In some cases, it was an external retractable mini-whip on the outset of the device's case; in other designs, it was a rubber-covered helical stub.

But several factors contributed to making antenna design both complex and challenging. With more and more functions and wireless technologies integrated in a single device, covering a multitude of spectra, the single-band monopole design was no longer suitable. In addition, cellular devices were shrinking in size, weight, and width even as more functionality was encased inside them. And consumers were attracted to cellular devices that looked good. It was a combination of function and fashion that sold such devices.

So, antenna designers were faced with a need to have the antenna inside a small device, packed with components, and the antenna system(s) had to support multiple technologies and spectra. To complicate things further, new health-related regulations more or less dictated where antennas would have to be located (near the edge of the device furthest away from the face and head); and directionality became an issue, too, so that radiation would have to be minimized toward the face and head.

Keep in mind that when a device's transmitter or transmitters are operating and antennas are radiating signals, those antennas are within fractions of an inch or inches from circuits susceptible to that radio energy. So, a lot of attention must also be paid to isolation of the antenna as well as directionality. Also, where multiple antennas are deployed, they must be isolated from one another to avoid mutual coupling. Where smartphones make use of MIMO technology, even if confined to the RX chain, this also complicates antenna design, placement, and isolation issues.

In today's smartphones, one finds the most common antenna designs are monopoles, for ungrounded antenna design, and planar inverted F antennas (PIFAs) for grounded antenna design. A lot of progress and innovation has been done for making these antennas multiband capable. For example, with PIFAs, one can shape the radiating path or use slotted ground planes to make these antennas efficient for different bands. Coupling of monopoles is another technique for multiband antenna design.

One can certainly play with dimensions and optimize antenna characteristics, but as with so many things, the variables from device to device can create big differences

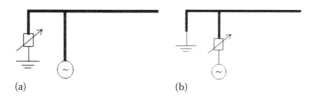

(a)   (b)

FIGURE 14.5   (a) Aperture tuning and (b) impedance tuning.

in results. Since the days of high-frequency (HF) radios and antennas, it was known that one could "tune" an antenna physically or electrically to obtain maximum radiated power, low-standing wave ratio, and the like. These principles apply equally to tweaking the antenna system of a smartphone for optimal results.

One can tune antennas in two fundamental ways: aperture tuning and impedance tuning (see Figure 14.5).

In aperture tuning, one may change antenna geometry, length, or grounding points that, in turn, affects both the antenna radiated efficiency (ARE) and its impedance. ARE is of great significance in antenna design because it is a metric that corresponds to the maximum total performance that an antenna can achieve, assuming that impedance mismatch loss is null. With impedance tuning, one conjugate matches the antenna to the transmitter power amplifier or receiver input impedance using some changes in capacitance and typically a fixed inductance to minimize the loss due to impedance mismatch.

Traditionally, one would do the impedance tuning by completing the following steps:

- Measuring the efficiency and impedance of the antenna
- Knowing the power amplifier's impedance
- Simulating by using the measured impedance data to find a matching network for optimum radiated performance

In Figure 14.6, one sees the application of a matching network for impedance tuning of an antenna.

In general, one will find it difficult, or impossible, to obtain a simple matching network that will provide a conjugate match for power amplifier and antenna for all the bands to be utilized, or even for all the frequencies within a band spectrum.

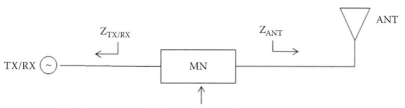

Matching network (MN) includes variable capacitors and switches

FIGURE 14.6   Topology for impedance tuning.

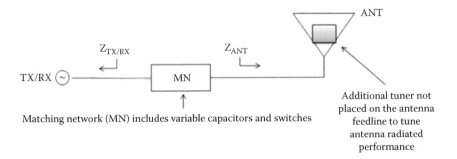

FIGURE 14.7   Topology for impedance tuning with aperture tuning.

Thus, a matching network, with multiple values of capacitances and a means of switching them in or out, on demand, can provide the match for each band and frequency as required.

Tuning an antenna for impedance matching reduces impedance mismatch but does not change an antenna's radiating efficiency (ARE). By using an aperture tuning method (see Figure 14.7), one can adjust both impedance and ARE. Of course, though, there is interaction in the settings and it takes much longer to adjust.

There are disadvantages in using a traditional method of tuning for antenna development:

- Doing optimization using simulations that must rely on accurate modeling and measurement of RF front-end impedance.
- Doing optimization using a passive fixture that, in turn, needs to have an RF coax cable so that power and impedance can be measured; thus, optimization may not lead to a true optimal performance in an active device.
- Taking long measurement time for conventional far-field measurement systems.
- Taking long development cycle for antenna tuning.

An alternative to the traditional tuning method for antenna development would be using an OTA system such as the one described earlier. It has the following characteristics, which are essential for this application: measurement repeatability, fast measurement time, and correlates well with far-field measurement system result. Thus, it lends itself well to OTA optimization rather than a complex methodology involving simulation, iterations, and long measurement times.

Figure 14.8 depicts an OTA system for real-time tuning by using a DUT's overall radiated performance rather than relying on impedance matching. As shown, the closed loop between tuning control and detected radiated power can be very fast.

The tuner control cable shown in Figure 14.8, however, may impose changes to results that differ once the device is no longer attached. Changes could include the cable's effect on the antenna performance characteristics, and the DUT receiver's sensitivity performance degradation due to noise generated from the external computer, which travels through the control cable to the DUT receiver.

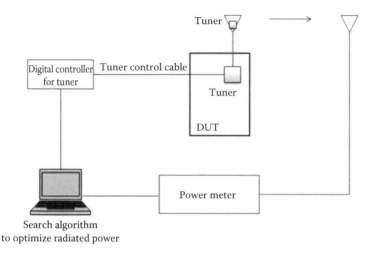

FIGURE 14.8 Example of a real-time antenna tuning OTA.

Figure 14.9, in contrast, shows a variation on this test approach where the tuner can be controlled by means that does not require any external cable connected to the DUT with the tuner. One approach for the computer to control the tuner is to use Wi-Fi or Bluetooth wireless connections. Another is to use signals from the tester that could switch different values of capacitance in the tuner while the OTA system provides fast power measurements. Measured performance, in this case, would be more consistent with the behavior of the DUT after the test since there would be no extraneous cables and connections.

Figure 14.10 shows an example of antenna impedance tuning, using a test setup as in Figure 14.9, where the measured metric is antenna efficiency, and not antenna impedance. A better impedance match will translate to better antenna efficiency. Results show overall antenna efficiency for 16 different capacitor values of a capacitor

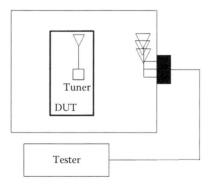

FIGURE 14.9 By eliminating any cabling between the DUT and tester during testing, measured results are more consistent with after-test findings than for methods involving cabling between DUT and tester.

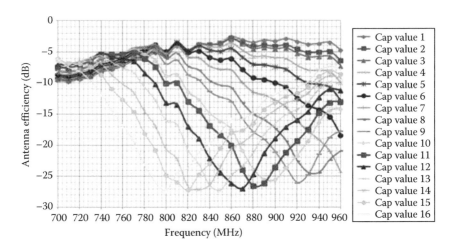

FIGURE 14.10   Antenna efficiency as a function of frequency for each capacitor value of a variable capacitor tuner.

in the tuner placed in a certain matching network configuration between the antenna and a transmitter for 27 frequency points from 700 to 960 MHz. With this plot, one can directly find the capacitor value that gives the best overall radiated performance for a frequency band. For example, one would choose the following capacitor values for the following bands: Cap value 13 for LTE band 17 (700–740 MHz), Cap value 9 for LTE B13 (740–790 MHz), Cap value 1 for LTE Band 5 and 8 (820–960 MHz).

As depicted in the plot, there are a lot of measurements, precisely 432 measurement values. Therefore, in order to utilize this method of optimization efficiently, measurement should be fast.

This method of antenna tuning over the air can be crucial in optimization time when antenna design includes additional aperture tuning where ARE is a varying metric. Impedance measurements cannot capture ARE. The only way to capture radiated efficiency variation from aperture tuning is to measure antenna efficiency. Therefore, OTA optimization on overall antenna efficiency that includes both radiated efficiency and impedance mismatch loss for antenna tuning makes a lot more sense.

The example shown here is for developing a device's antenna for covering the low-band spectrum while maintaining adequate antenna efficiencies. In a case where a device has a built-in matching network and where the OTA test system can control that network's switching while measuring antenna performance, it would be possible to fine-tune each device's antenna performance during the fully assembled final test phase.

14.7   CONCLUSIONS

Testing of wireless devices after manufacturing is essential to detecting manufacturing defects that render the device inoperable or out of specification. This is important because one does not want to ship a faulty device or one whose operation may cause interference to other devices and services.

Manufacturing testing is driven by the PHY layer specifications of the standards that underlie each technology, such as 802.11n or 3GPP LTE. Manufacturing test should not be used to verify a design. That is the purpose of device design verification testing. Rather, manufacturing test must assume a design is valid and look for indications that a device, whose design is sound, has been affected by manufacturing defects. This allows for using a subset of design verification tests for faster and lower cost testing.

Contemporary manufacturing test of wireless devices, ironically, makes copious use of signals conveyed conductively rather than radiated. As such, because the test systems must connect to a device's antenna port(s), the device cannot be fully assembled and intact. In fact, one cannot test the entire device because the antenna will not be connected during conductive testing.

Consequently, fully assembled devices are tested, again, later to ensure that the device works in an OTA signal environment. Today, the proportion of tests and test time done conductively versus OTA is both richer and longer. However, innovations in OTA testing systems and methods are paving the way for more tests to be done using OTA methods. Whereas the actual testing of TX and RX characteristics may not change compared to conductive testing, the handling time could certainly be shortened by the absence of connecting and disconnecting conductive cables to devices' antenna ports.

Index